Farms of Tomorrow Revisited

Community Supported Farms
Farm Supported Communities

Trauger Groh
and
Steven McFadden

Biodynamic Farming and Gardening Association
Kimberton PA

The body text of this book is set in Adobe Garamond and Adobe
Garamond Expert; the titles and heads are in Adobe's version of Post
Antiqua Medium.
Design, layout, composition and production by Bruce Bumbarger.
Cover illustration by Elizabeth Auer.
Printed by Thomson-Shore, Inc.

ISBN: 0–938250–13–2
Printed in the United States of America

First Edition

To Bernhard Hack
(died January 5, 1997)
and
Robyn Van En
(died January 8, 1997)

Table Of Contents

Preface

Our book, *Farms of Tomorrow: Community Supported Farms, Farm Supported Communities,* appeared in print seven years ago with the help of the Biodynamic Farming and Gardening Association. It contained basic essays on new structures for farms which acknowledged that farming is not just a business like any other profit-making business, but a precondition of all human life on earth, and a precondition of all economic activity.

As such, farming is everyone's responsibility, and has likewise to be accessible for everyone. This fact makes new social structures for farming necessary. Such new farms, known as community farms or CSAs, started as simple, isolated "test plots" in the 1970s in Japan and Europe, and arrived in the mid-1980s in the United States.

In addition to basic essays, *Farms of Tomorrow* contained case descriptions of beginning community farms to give farmers and those who aspire to become farmers practical advice, and models of how the new ideas could be brought to life. The book was very well received in the United States, and in many other countries because it seemed to meet an imminent need in a world situation where fewer and fewer people are participating in farming, which has become largely industrialized and mineralized, with negative consequences for food quality, life quality, and the environment.

Over time *Farms of Tomorrow* appeared in translation in South Korea, Russia, and Japan. When the original English-language version was sold out, the question arose of whether to reprint that version, or whether the experience of nearly seven years could justify a new book. We decided for the latter—to maintain the core of the first book, and to add the insights we and many community farmers have gained over the years. We have named this new book *Farms of Tomorrow Revisited,* for through it we revisit both the concepts, and the actual farms described in the first book, and are thus able to document their development. Farms that have been given up in the meantime have been omitted; and new farms that reveal new aspects of the central questions have been added.

In addition, six new essays concerning the community farm movement were written and appear in this new volume. Grateful we are for the generous and worthy contributions of Example 6 by Sam Smith; Example 7, aided by members of the Roxbury Farm CSA; Example 8 by Annie Main; Example 9 by Jim Brons; Essay 4 in Part 2 on community farm coalitions by Marcia Ostrom; and Appendix B on community land trusts by Chuck Matthei. Grateful we are again to those many people who contributed descriptions of the community farms, or who contributed their insights via discussions or shared materials.

As we were completing work on the manuscript for *Farms of Tomorrow Revisited,* Bernhard Hack died at his farm on January 5, 1997. After having established a perfect model of the biodynamic farm in Germany, and again with his sons after 1982 at Lake Huron in Ontario, Canada, Bernhard took on the task of giving spiritual and practical advice to farmers in Ontario, Quebec, and, over the last seven years in Russia. His restless, selfless efforts for the new farm on two continents, his enthusiastic dedication for the biodynamic farm concept, exhausted finally his life forces. He will be unforgotten by hundreds of grateful farmers.

Just three days after Bernhard's death, on January 8, 1997 in Massachusetts, Robyn Van En moved in a tragic way through the gate of death. Robyn was a tireless pioneer of the CSA movement. She did not agree with all the positions of the authors on agriculture and community farming, but her active promotion of new ways of farming, and the enormous help she gave to farmers who were looking for new ways, should not be forgotten. Thus, we have dedicated this new book to both Bernhard and Robyn, in remembrance of the wonderful contributions they made through their lives.

We hope that this new book, written out of practical experience with the emerging farms and farmers of the future, may be so well received as its forerunner, *Farms of Tomorrow,* and that it may serve as well both farmers and communities.

—Trauger Groh and Steven McFadden
January, 1997

Acknowledgements

This author has had the good fortune to be included in the serious striving for the farms of tomorrow taking place in both the old and new worlds. If not for this good fortune, this book would not have been written, and some of the farms discussed here could not have been established. Some of the farmers involved in this striving I am proud to call my friends. They have struggled by themselves and in cooperation with others to develop new concepts. They have been ready to take great risks trying new ways with their land, with their labor, and in addressing the basic questions of farming that are described in this book. Some of these farmers were very much alone in this striving at first, but they have held onto their ideals to develop a greater understanding for new methods.

Two close friends of this author were particularly helpful: Heiloh Loss and Carl-August Loss. They decided in 1978 to donate the only property they had, a 200-acre farm that carried practically no mortgage, to a land trust. This action made possible an innovative experiment in cooperation with the author and with other people on this free land. The new farm they created together, Buschberg-Hof, has proved that free individuals can cooperate in a farm operation that works land held in trust, rather than as private property. This experiment also established the possibility that a farm can succeed when it strives for quality rather than profit, and that by this the farmers can support themselves and their families more securely than if they had been running the farm for profit.

In addition to the farmers who helped this author put new ideas into practice, persons have to be mentioned who gave him impulses and enthusiasm out of their life experience in the field of agriculture and social economics. Dr. Nicholas Remer, a farmer and agriculturist who worked as a researcher and farm adviser with hundreds of farmers over fifty years, allowed the author to take part in his spiritual striving as an agriculturist, and shared his vast experience. He offered his help in many ways in establishing ecologically sound farm organisms based on a new social concept. Dr. Remer belongs to the group of farmers and scientists who developed the biodynamic approach to agriculture through practical application and continuous research. As a result of these pioneering efforts, we can say today: biodynamics rightly applied can provide mankind with ample healthy life-giving food, and can not only conserve our environment but improve it where it has been neglected or abused. Two of his books are available in English: *Laws of Life in Agriculture* (Biodynamic Farming and Gardening Association, 1997), and *Organic Manure and its Handling, Following the Advice of Rudolf Steiner* (Mercury Press, 1997).

This author is also deeply indebted to Wilhelm Ernst Barkhoff, a lawyer and economist who has taken a deep interest in the socio-economic problems of farming. He gave me and many other farmers the inspiration to rethink legal, social, and economic conditions of agriculture, and helped us to set up models where people could cooperate agriculturally on free land, including even non-farmers who took an interest in this task. I learned from him to leave worn-out tracks of thinking and start a fresh approach in an unprepossessed way. Many years ago he gave me three golden rules for community farming that have proved precious and valid for any such endeavor. I am pleased to present those rules in Part I of this book.

Finally, I want to express my gratitude to Alice Bennett-Groh, Ph.D. She carried on all the consequences of family life that arise out of the fact that I am away from home learning, lecturing and advising on these matters, besides writing manuscripts and participating actively in the Temple-Wilton Community Farm. Beyond that she is ready to discuss every thought and every measure that goes with this activity in an understanding and helpful way.

It is just not possible to acknowledge the hundreds of books and articles that I have read on farming and related matters. Yet I realize and acknowledge that many sources have influenced and helped me to shape the picture I have tried to present in this book. For all of those writers and their many insights, I am deeply grateful.

—Trauger Groh

My earliest interest in the garden arose from observations of my mother's efforts with a bed of reluctant peonies. Though I also was reluctant, she drafted me as a child laborer. Somehow, through sharing with her an intense and oftentimes frustrating interest in that bed of flowers, and also the love of mother and child, within me was planted the seed that in later life blossomed into a consuming interest in gardening and farming.

My primary teachers have been the plants themselves, especially the flowers with their many soul-healing attributes. And so to these realms I give deep thanks.

I also wish to acknowledge my neighbors, the farmers of the Temple-Wilton Community Farm: Lincoln Geiger, Anthony Graham, and Trauger Groh. Their deep thought, their beautifully coordinated efforts, and the impressive results of their work, have steadily inspired me to advance my own cultivation practices. Much more lies ahead.

—Steven McFadden

The co-authors wish to acknowledge and extend their gratitude to the Biodynamic Farming and Gardening Association for making this book possible, in particular executive director Charles Beedy and his associate, Jean Yeager. This book has been greatly enriched by the insights of the many farmers and others who spoke with us, and shared their time and experiences. In a certain sense, through the contribution of their views, the creation of this book has been a "community supported writing project."

Introduction

"When tillage begins, other arts follow. The farmers therefore are the founders of civilization."—Daniel Webster, New Hampshire statesman.

Agriculture is the foundation of modern civilization. Without a steady supply of clean, life-giving food, we have neither the leisure nor the energy to develop industry, science or art. Worldwide, and in particular in the United States, our foundation has deteriorated dangerously. It requires immediate and fundamental restructuring. But how can we even begin to approach this task?

This book has been written to suggest some possibilities, and also to serve a need that is becoming more and more explicit: the need to share the experience of farming with everyone who understands that our relationship with nature and the ways that we use the land will determine the future of the earth. The problems of agriculture and the environment belong not just to a small minority of active farmers; they are the problems of all humanity, and thousands of people are searching for new ways and new solutions.

As the farming crisis deepens, many people are seeking wiser, more effective ways to reestablish the relationship of human beings with the earth. The financial and agricultural practices of recent decades have made it increasingly difficult, and in some cases impossible, for existing models of agriculture to prosper. In America, the family farm has fallen victim to a relentless marketplace; meanwhile, corporate farms have tended to place short-run economic advantage over the long-term considerations of our relationship with each other and the earth. Modern ways of industrial and chemical farming play a major part in the deterioration of our environment on all levels: soil, water, air, landscape, and plant and animal life. Only a new, ecologically sound approach to farming can slow down or stop this deterioration.

When we look to the universal questions of land use and land abuse, and see the manifold dimensions of these questions, we understand quickly that there is no universal solution. There is no simple recipe or remedy for the many challenges we face. Out of this understanding, the authors decided upon the following approach: first, to work out some fundamental questions and principles of land use as it concerns our food, our environment, and our general ways of living with the land. Second, to present living examples of a new approach to the use of land. And third, to offer readers a list of resources so that they may have ready access to information which will support them in the pursuit of new, healthier uses of the land.

Part I was written by Trauger Groh, and it represents the fruit of thirty years of experience in practical farming and advisory work, as well as numerous lectures in various countries of both the new and the old world. Parts II and III were primarily written by Steven McFadden, a journalist with a special interest in ecological and agricultural questions. The resources at the end of the book were gathered by Steven, and by Rod Shouldice, former director of the Biodynamic Association.

The experiments in farming described in this book represent new social forms of agriculture which have arisen in recent years while traditional family farms have declined and industrial agriculture has increased. These new farms involve many local families directly in the decisions and labor which produce the vegetables, fruits, milk, and meat they eat. In that way they reestablish a link between the farm, the farmer, and the consumer. While this approach may not be the full answer to the questions posed by the modern agricultural dilemma, we believe it has much to offer.

In simple terms, these efforts arise under the name Community Supported Agriculture (CSA). A CSA is a community-based organization of growers and consumers. The consumer households live independently, but agree to provide direct, up-front support for the local growers who produce their food. The growers agree to do their best to provide a sufficient quantity and quality of food to meet the needs and expectations of the consumers. In this way the farms and families form a network of mutual support. Within this general framework there is wide latitude for variation, depending on the resources and desires of the participants. No two community farms are entirely alike. The authors tried to select examples that show great variety in approaches to the farms of tomorrow.

As with many catchall names, the term community supported agriculture or CSA is slightly misleading. It implies that the problem is special support for agriculture. As important and necessary as that may be, it is secondary. Although it may seem a fine point, the primary need is not for the farm to be supported by the community, but rather for the community to support itself through farming. This is an essential of existence, not a matter of convenience. We have no choice about whether to farm or not, as we have a choice about whether to produce TV sets or not. So we have to either farm or to support farmers, every one of us, at any cost. We cannot give it up because it is inconvenient or unprofitable.

Since our existence is primarily dependent on farming, we cannot entrust this essential activity solely to the farming population—less than 1.9 percent of Americans. As farming becomes more and more remote from the life of the average person, it becomes less and less able to provide us with clean, healthy, life-giving food or a clean, healthy, life-giving environment. A small minority of farmers, laden with debt and overburdened with responsibility, cannot possibly meet the needs of all the people.

More and more people are coming to recognize this, and they are becoming ready to share agricultural responsibilities with the active farmers. Out of this impulse, up to sixty CSAs developed in America from 1985 to 1990, and many hundreds more from 1990 to 1997. From these, the authors have selected several different farms as models. In recognition that the deterioration of our farm system is frequently caused by our money system, we have made a special effort to explore new ways of farm financing.

Some things are typical for all community supported farms. In all of them there is a strong dedication to quality; most of them are organic or biodynamic farms, most of them show great diversification, most are integrated farm organisms having their own livestock and thus their own source of manure, or they are aiming in this direction. At all of them, far more people are working regularly per hundred acres than in conventionally run farms; and generally there are just many more people around participating in all the dimensions of agricultural life: working, relaxing, storing, shopping, celebrating. This human element is of enormous importance. It shows that these farms have something to offer beyond good food. They embody educational and cultural elements that draw the interest of many people. Besides clean, healthy, life-giving food, and a strong contribution to an improved environment, the educational and cultural elements constitute the third great gift that the farms of tomorrow have to offer.

Neither the urban nor the suburban life of today is sufficient for healthy, continued human development. In the future, to give the soul-nourishing experience that men and women require—especially children and teenagers—they must have the opportunity to engage actively and creatively with life-filled farm organisms.

The growth of Community Supported Agriculture has been surprisingly swift. As of 1997 it is estimated that there are an estimated 1,000 CSAs in the United States, involving perhaps 100,000 households, and many more similar farms in other nations. Community farming, however, is still at the seedling stage of development.

Over the course of experience we have learned that while community farms confront a host of challenges and questions, they do work: they feed people, they save energy and money, they take care of the land, and they bring networks of independent households back into direct connection with each other and the Earth. For increasingly large circles of families and farmers, community farms are working. If not growing all the answers, these farms can at least be said to be cultivating the right questions.

Because of poor growing technique, poor financial management, or poor community building skills, many CSAs have risen only to collapse. Yet more community farms succeed than fail, and many important lessons have been learned. Through their collective experience, CSA farmers and shareholders have learned the typical pitfalls, and also the parts of the undertaking that really work. We have striven to include those lessons in *Farms of Tomorrow Revisited.*

The concept of associative economy underlies much of the community farming movement. This concept is explored in two places in particular: in Trauger's Essay 5 on the *Economic, Legal, and Spiritual Questions Facing Community Farms,* and also in the example of the Hawthorne Valley Farm. Readers will find guidance toward broadening their understanding of this concept in the *Bibliography* and in the materials listed in *Appendix F—Resources.*

As we conceive of it—and as it is being practiced and developed at a great many farms—CSA is not just another new and clever approach to marketing for farmers, even though some people have chosen to regard it that way. Rather, community farming is about the necessary renewal of agriculture through its healthy linkage with the human community that depends upon farms and farmers for survival. From experience, we also see the potential of community farming as the basis for a renewal of the human relationship with the earth.

Essays on the
Farms of Tomorrow

Trauger Groh

Essay 1

Why Do We Need New Farms?
Food–Environment–Education

When we speak about the need for healthy farm organisms, we think first of our food supply and then we think of the farm as part of our natural world, shaping the environment in positive or negative ways. Rarely do we have in mind the great contribution that living on farms and working in nature gives to our inner soul development and to the shaping of our social faculties. Yet all these considerations are essential elements of agriculture, and of the farms of tomorrow.

Healthy Food

The question of food and food quality is very complex. We speak in general terms about healthy food, or life-giving food. But these terms can mean different things to different people. In the modern context, perhaps a more accessible concept is "clean food," clean meaning free of any synthetic substances that might be added during growing, processing or preserving. Such substances are typically preservatives, insecticides, fungicides, herbicides, synthetic colors, and so forth. Arguments about which additives are tolerable and which pose a health threat are complex and confusing to most people, and so they let the government step in to make such determinations. But we should be skeptical towards authorities who decide these questions for us. It is extremely time-consuming and difficult to establish the exact health effect of any of the many synthetic substances that are routinely added to our food. One thing we can say with certainty. The cumulative effect of the different substances that are added is largely unknown. Government agencies such as the Food and Drug Administration (FDA) are simply not in a position to guarantee the safety of any food additive, even if they pretend to be. They are not even able to test properly what is in use, never mind the new synthetic substances constantly being introduced to the market. The standard declaration of additives on food packaging is a good thing, but the

Reprinted from the original *Farms Of Tomorrow*

widespread belief that what is declared, and therefore allowed by the government, is without problems is an illusion.

That leaves the wise consumer only one choice: to demand food without any additives. If we ask for a carrot, we should demand carrot, and only what nature gives us in the carrot. If we ask for milk we should demand milk in the beautiful composition given by a properly fed cow, not accepting anything more or less, such as the synthetic vitamins typically added during processing, or the loss of the life essence that occurs during pasteurization.

Can we, or should we, allow the state to make the basic decisions about what is good for us? Is this not a basic right? In the case of milk, for example, the government has assumed the right to decide that milk as nature gives it is hazardous to human health, and that therefore all milk must be heat processed in a way that changes markedly its natural composition, robbing it of essential parts and driving out all the life forces that are in it. If someone wants to consume raw milk or some other forbidden food, and if that person believes the food is good, and also has a trusting relationship with the farmer who produces the food, should they not have that right?

The absence or presence of additives alone does not determine the quality of food. The fundamental secret of quality production is to handle the plants and animals so that they attain their highest performance by their own nature. In each creation, there is an inner harmony of substances and forces that is typical and healthy. It is not the presence of certain substances in certain amounts that makes a vegetable or grain healthy; rather, it is the harmonious relationship between the substances and the forces. To a large extent, modern agricultural methods have drastically affected this harmony. As research has shown, between 1896 and 1932 many crops exhibited a strong rise in the content of potash while their magnesium content declined. Meanwhile, other research shows that the silica content in cultivated plants has tended to decline while the potash content has been rising.[1]

The results of this change to a less harmonious balance showed up in Eastern Europe, where for hundreds of years people thatched their roofs with rye straw. Those roofs typically lasted for fifteen years. But after the rye crops were treated with synthetic nitrogen, and the natural harmony of substances and forces had been altered, the roofs fashioned from the resulting straw began to rot after just three to five years. Though perhaps not so obvious, similar changes have occurred in the bread grain that is a staple of our diet. There the weakening of the plants through inharmonious fertilization shows up in the excessive appearance of fungus diseases, which again provokes the use of harsh fungicides. As for the grain itself, the potash and phosphorous content is higher today than 100 years ago, and the silica content is less. What influence does this profound change have on the human beings who eat the bread and other products made from this grain? Some observers believe the high phosphorous content in many processed foods, much of which comes through industrial food processing, is a major factor in problems of hyperactive children, and other observers believe that the reduced silica content has led to a dulling of our senses.

While science has developed highly sophisticated ways of making quantitative measurements, the concept of quality is difficult to measure with gauges and scales. To evaluate quality, we must observe how the food affects the higher organisms who digest it. For example, carefully designed tests have conclusively demonstrated the effect of organic, bio-dynamic, and conventionally grown grains upon the urine of domestic animals. If the quality of the food can be detected in the excretion of an organism, then clearly the quality of the food is having an affect on the health of the organism.

As we create the farms and the culture of tomorrow, we need to aim in a certain direction with our nutrition. What do we want to achieve with nutrition besides keeping up our bodily functions? How can our diet support not only our physical health but also the development of our spiritual faculties so that they function in the best way? The point that men and women live longer today than in the past is a poor argument for the quality of our food if we do not pose questions about the condition of our lives. What do we want to achieve in life? Are we really in full possession of our faculties of thinking, feeling, and willing? One striking example of dulled spiritual faculties comes in the realm of free will. In general, modern men and women have strong will forces, but their lives lack direction and creativity; the will forces are not channeled in the service of creativity. Common deficiencies in enacting ones will forces and the moral insanity that we perceive all around us may well be connected to the low quality of the food generally available for consumption.

Food quality is first determined upon the farm by the way we interact with nature and its forces. The profit motivation does not lead to quality food production. This thesis can be proved by looking into the history of modern farming in the last 100 years and into the state of affairs with our processed foods. Farming differs here from the production and marketing of industrial goods. You cannot sell cars that have grave deficiencies for very long, but you can deceive mankind for a long time with deficient food. The consequences of a deficient car show up very rapidly, but the effects of deficient food—nicely colored and flavored with artificial ingredients—are much harder to discern, and turn up mainly in the soul life of humanity or in the health problems of old age.

Nearly all manipulations with food—additives, radiation, and conservation methods—serve not the purpose of quality, but rather the purpose of distribution over long distances, shelf-life, and a pleasing appearance. Contrary to what might be right for many industrial products, the production, processing, distribution, and consumption of food favors quality when it is done locally. At the same time, this is the most economic approach to food because it saves transportation and preservation costs. The community supported farm systems of the future will proceed in this way, that is, producing for the local community, which includes the closest cities. Here households will connect themselves with local farms directly or via trusted agents so that they can support a system of production that aims primarily at quality rather than profit.

A Healthy Environment

Protection of the earth, the air, and the water, and the positive development of an already desecrated environment is not possible without healthy farm organisms. This becomes obvious, for example, in those parts of New England where farming has ceased to exist, and where the landscape consists only of overgrown farmland and suburban development. Once the farms with their animals are gone, and with them the open fields and pastures, the scenery loses its soul-animating quality. On the other hand, we can see contemporary industrial farms as one of the great destroyers of the environment. The vast ground water pollution of our days, the extreme erosion of topsoil, these are just some of the consequences of modern, profit-oriented farm systems. As they are established, the farms of tomorrow must protect against the following abuses:

- The soil against one-sided agricultural uses, and the poisonous effects of artificial fertilizer, pesticides, herbicides and fungicides.
- The water against pollution, lowering of the water table, and the consequences of soil erosion.
- The air against poisonous emissions.
- The life of wild plants and wild animals against the dwindling of natural habitats, leading to eventual extinction.
- The cultivated plants and domesticated animals against degenera-

tion and exploitative management practices, including biological engineering.

- The landscape against monotony and sterility.

All this we can achieve only with the help of ecologically sound farm organisms that, while producing the necessary food for humanity, not only protect the environment but also cultivate the land in a healthy way.

Ecologically sound farm organisms mean a new type of farm in which there is created and upheld a balanced relationship between animal husbandry, acreage, field and pasture on one hand, and forest, hedgerow, water, and fallow land on the other.

If we envisage this type of farm—and we have some good examples among the community supported farms described in this book—and we compare them with the monotony of modern farm industry, then we become aware what the farms of the future will mean for our well being and for the future of planet earth.

The Education of Humanity

Besides quality food production and environmental care, there is a third essential task which the farms of today in their one-sided industrial production have tended to neglect, and which the farms of the future must take up again. That task is the education of humanity through active work in nature, specifically nature that is formed into healthy, self-supporting, ecologically sound farm organisms. If we look back into the agricultural societies of the past, and this past lies only 200 to 300 years back, we find a very simple school system. For those children who went to school at all, it was thought sufficient if they learned the stories of the holy books, to read and to write, and perhaps some mathematics. But for most people, the great educators of olden times were the farm and nature. The farms of tomorrow can also teach many profound lessons, within relation to the technological context of modern times, including the following:

- A rhythmic lifestyle shaped by the seasons and the rhythms of the day—that is, the rhythms of the sun. In the modern world we have been liberated from natural rhythms through electric lights and other appliances. But people who use artificial light to completely reverse day and night, tend to become ill. They are working against the rhythm of nature, rather than with it. A life lived in accordance with natural rhythms will include, for example, natural pauses for observation and reflection at the Solstices and Equinoxes, the turning points of the year.

- A modest lifestyle adapted to what nature locally, with the help of farmers, can give. When we are in touch with the capacities and limitations of nature, this awareness spills over into other dimensions of life and helps establish balance. By contrast, in the United States where relatively few people are in contact with farms or nature, there is a feeling that everything is possible if there is enough money. The result of this attitude is a nation that consumes 50 percent of the world's resources, and is choking on its own waste as the landfills overflow. The farms of tomorrow can naturally instruct humanity in a lifestyle that is not only more modest, but also more satisfying.

- A readiness to do what is necessary without complaint—feeding and cleaning the animals, caring for and harvesting the crops, and sublimating one's pleasures to these duties. Unlike the demands of parents or teachers, the demands of the farm can be understood and accepted by young people. When a parent demands that a child behave, or do homework, those demands can seem abstract. But when corn must be harvested, or sheep fed, the demands are tangible, and this teaches the value of work and of service.

- A deep feeling for the self-evident fact that you cannot harvest anything without planting. This is fundamental logic, in concrete rather than abstract terms.

- An appreciation that in reality it is nature that produces on the farm, not the farmer, and that natural production on the farm is not an input-output equation, but rather a cooperative venture with the forces of the earth and the cosmos. The only new wealth in the world comes from the forests and the fields, as each year nature renews itself. As it stands now, farming is a wasteful enterprise, for the input of substance and energy is often higher than the output. This is inherently unbalanced and is leading to many deep problems. Such an understanding has enormous importance for our economic world view.

- An understanding that working on farms means relying on what other generations have done—clearing the land, draining the swamp, picking the stones out, and so forth. From this understanding arises the recognition that your own contribution will not necessarily serve you and your family alone, but also future generations. If you plant a white oak or a redwood tree, the time of its maturity will arrive long after you and your children have passed on.

- That nature produces best out of the great variety of plants and animals, not out of one-sided systems, or monoculture.

• That nature needs animals and animal manures to keep the fertility of the soil.
• Hundreds of technical skills necessary for skillful use of tools and machines.

All these basic teachings our educational systems and our mostly urban and suburban environment do not provide for modern children. Most children grow up in a way that does not acknowledge or harmonize with the rhythms of nature, or even the rhythm of their own bodies. They grow up in a way that ultimately convinces them that money brings good, and that pleasure in and of itself will bring happiness rather than the work we do for others. Out of this education, they follow their uncontrolled desires, rather than responding intelligently to the necessities of a natural environment. They think that everything is available if only one has the money to buy it, and that the world has never-ending resources. They often see in their surroundings that people live well without contributing to society. They cannot experience that the critical difference between industry and agriculture lies in the fact that nature can produce out of its harmony and variety without major input, while industry cannot.

In these conditions young people miss the basic social experience that comes from recognition that in cultivating the earth and caring for animals and plants, one must rely on the work of others who cultivated before you, and that you do not necessarily reap what you have planted, but that others may benefit from your work.

Can the farms of tomorrow again, and in a new way, provide these teachings for young people? If we see the disastrous state of our educational systems now, with their overemphasis on intellectual faculties and their incapability of helping young people to create an inner morality, of directing their will forces creatively out of a strongly developed personality, we can recognize that we have to do everything to bring the farms of the future into a condition where they can directly contribute to the inner and outer development of young people.

We cannot move back to a rural society. We have to create a new relationship between the citizens and "their" farmland that will make the benefits of farm experience available for anyone who seeks education, recreation, or therapy. Every community needs to incorporate farms not only to have fresh local food, but also to have available these educational facilities. The extermination of farms on the East and West coasts of the United States is leaving a vast, thinly populated, highly mechanized and chemicalized agriculture in between. This has many detrimental consequences, environmental, educational, and social.

Every school, public or private, needs a farm or a group of farms to give students the opportunity for practical training in nature. Therapeutic communities with handicapped, retarded, and chronically diseased people and with juvenile delinquents should also be placed inside or adjacent to healthy farm organisms, and make use of them. Likewise, elders should have the opportunity to work in gardens or farms. In fact, to compensate for the soul-draining work of factories and offices, such opportunities should be available to all.

This will become possible only if farms are diversified, integrated, and oriented toward quality instead of profit. Many existing biodynamic farms have long experience with such educational and restorative efforts, serving as the haven for whole classes of retarded and elderly people, as well as young families with small children who just come to look and play. Naturally, many implications and difficulties can arise out of such situations, but they are resolvable.

For the farms of tomorrow to undertake this broad educational and therapeutic task, they will require new generations of farmers who have not only mastered the farm techniques but who are also strong in humanities and social skills. Healthy food, environmental care, and educational possibilities—for all these essential sources of life and for inner development, we need the new farms.

1. Nicholas Remer, *Lebensgestze im Landbau,* (translated as *Laws of Life in Agriculture.* Kimberton, PA: Biodynamic Farming and Gardening Association, 1995); Gerhard Schmidt, *The Dynamics of Nutrition.* Kimberton, PA: Biodynamic Farming and Gardening Association, 1987.

What Is Needed to Create the Farms of Tomorrow?
Concept–Land–People

The farms of tomorrow must arise from a new concept, a new leading idea that serves the basic aims of agriculture. Those aims are, first, to grow life-filled, health-giving food in ample quantity and diversity to feed the local community and to serve regional and urban needs that are not met locally; second, to do this in a way that not only conserves but improves the natural environment; and third, to give all who want it the educational experience of working with nature. Without a leading concept that concerns itself with the wisdom that lies in nature and with the relationship of the human being to nature, we will be unable to create new farms that will serve these three purposes.

In the past, the motivation for agriculture was primarily taken from the need to support oneself and one's family with food, firewood, and clothing. The methods of farming were shaped by experience and the traditions that resulted from them. Far into the eighteenth century, farming was not so much an economic venture as a means of self-support, and also the general lifestyle. In that sense, it was pre-economic. Before industrialization and the growth of cities, most people were engaged in farming. There was no real market for agricultural goods. For many centuries the only money that was needed was money to pay taxes to support the nobility, their soldiers, and the clergy who did not support themselves through farming, and also to buy necessities such as tools for farming and salt. Salt was essential because in cold climates one could not survive the winter without salted meat, fish and vegetables. To get the little necessary cash for these things, many people went into a craft or a service business without giving up farming. They became blacksmiths, carpenters, or innkeepers in their home villages, and by this created tradable goods or services. So the general pattern was for rural people to keep farms to feed themselves, and produce goods or render services to trade. Only toward the end of the eighteenth century did farming, very slowly, become a business itself. It was in this epoch that agronomists like the German Albrecht Thaer proclaimed "agriculture is a trade, the purpose of which is to make profits or money. Farming is a way to earn money like any other business."

The motive to earn money through farming, to make a profit—profit being the difference between money input and money earned—took its place beside the traditional values of farming, and steadily became more and more domineering. The rapid development of natural science in the eighteenth and nineteenth centuries, and the concurrent development of agricultural science, provided the tools for a vast and necessary expansion of agricultural production. Modern agriculture was formed through the combination of the new economic approach and agricultural science with the rapid growth of population and the expanded economic resources available through industrialization.

Agricultural science took more and more to the new economic trend. It aimed less at exploring the ideal conditions under which a whole farm with its plants and animals thrives as a natural organism. Instead, science turned the art of agriculture into agronomy, techniques of exploiting soils, plants and animals for monetary profit. The guiding question of agricultural science has been, under what conditions is plant or animal production the most profitable—with profit measured solely in money. The nature of the farm organism and the question of its relationship to the environment was rarely considered.

Generally questions of quality became, and still are, secondary to questions of profit. If we look at the farm scene of America today, at the farm crises of this century, at the devastating impact of this approach to farming and to our natural environment with its vanishing soil, its sick and vanishing forests, its polluted ground water and its often miserable rural population, we perceive what a high price not only the rural population, but the whole of society has to pay. It has become obvious that the profit motivation does not lead to healthy life-giving food, nor to conservation or improvement of the environment. The history of agriculture in the last 200 years proves this clearly.

As stated in the first essay, we need farms for three reasons: for healthy food, for a healthy environment, and for cultural and educational reasons. In dealing with these needs we have to be aware that they are basic to everyone, and in creating the farms of the future we have to make sure that the needs of all are met. Consequently, three different motivations have to come together to shape the farms of tomorrow.

- The first is the basic spiritual motivation: that every year life on earth is created anew, so that human beings can be born safely and have healthy bodies that will allow them to live out their individual and collective spiritual destinies.
- The second is a social motivation: to shape our land use with the goal that everyone have access to healthy food, wood, and fiber in the right amount and independent of his or her life situation.
- The third is the economic motivation that makes all other goals possible, and is the basis of the new farm concept. We must develop the farms of tomorrow in such a way that they regenerate themselves more economically and become more and more diversified, serving as the primary source of food for the local community. This diversity and regeneration should arise with the help of the forces of nature inside the farm organism so that it becomes less and less necessary to introduce into the organism substances and energy from outside such as feed, manures, and fuels, and so that human labor is used as economically as possible. Stated another way, the economic ideal is a farm that achieves and maintains high fertility within itself, generating a surplus of food for the community, and its own seeds for the coming year while the input of outside substances, energies, and labor goes toward zero.

This truly economic motivation should not be confused in any way with the profit motivation. They are totally different categories. Many things we do today in farming are profitable but uneconomic. For example, nowadays strawberries are frequently grown in California for the Northeast market, even when those berries could be grown in the Northeast itself. The grower, the trucker, and the retailer will all eventually make a profit, but there is a hidden loss. The amount of energy expended to grow and transport the strawberries to market far exceeds the amount of energy they will yield when they are consumed. There is a far higher input than output. Ultimately, society must cover this loss in some way. Thus, the profit of a few becomes the loss of many. Production is truly economic when it is done with the lowest possible input of substances, energy, and labor, and when the output exceeds the input. As we enter an era of dwindling resources, many people are recognizing the need for the farms of tomorrow to be truly economic: to renew themselves while creating a consumable surplus without the input of substances from off the farm, and with a reasonable expenditure of labor and energy.

A Leading Concept

There is a well-developed agricultural concept that can guide us as we work to create the farms of tomorrow. This leading concept was given to European farmers in 1924 in eight lectures by Dr. Rudolf Steiner, the Austrian philosopher and spiritual-natural scientist. This concept has come to be known as biodynamic farming. It has been put into practice over the last sixty-five years on hundreds of farms on all the continents. In America it took hold mainly through the work of Dr. Ehrenfried Pfeiffer. Biodynamic farming serves predominantly the motivations mentioned above. Over the years it has reliably demonstrated its capacity to lead farms to high quality on all levels, and to the most developed economy.

The idea of the farm as organism is widely unknown today, yet it is of high importance for those farms that work with the advice of Dr. Rudolf Steiner. In the biodynamic approach, the farm is seen as an organism, and that underlying concept is part of all considerations and actions. By definition, an organism is a living entity consisting of parts, or organs. The function of each part is essential to the existence of the whole, and also to each of the parts. An organism has its own inner life and circulation that is different from its surroundings. An organism develops over time, having a beginning and an end, and its performance depends on the harmony of its parts, or organs.

More and more scientists are beginning to perceive the whole earth as an organism, thanks in part to the work of Dr. James Lovelock and the inspiration of his Gaia Hypothesis. Because of the essential exchange processes of oxygen and carbon dioxide, the earth's destiny depends on a harmonious ratio between mankind with the animal world and the plant world. As everybody can see now, we need the right ratio between forest and open land; we must have, for example, the oxygen-producing and water-conserving belt of rain forests.

In this context arises the question, how can a farm be or become an organism in a true sense? Every farm and every organism lives in space and has boundaries. There are legal boundaries, such as the title, which defines the land which can be part of the farm; but legal boundaries are abstract and create no living organism. In nature, space is created and formed by the animals. We are living in a network of animal territories. Wild animals mark and defend their territories; naturally these territories are not related to our legal boundaries. Deer, for example, move over a large territory not observing any legal boundaries.

The organism farm is created by the domestic animals. Inside the farm's legal boundaries the farmer establishes a herd or various herds.

Ideally, they feed from the vegetation inside the farm, summer and winter, without anything brought in from other farms. The animals respond with a manure that is formed exclusively by the flora of the farm organism. This manure is collected, and when properly treated comes back to the plants of this place, stimulating them. In this process of correspondence between farm animal and farm vegetation, the farm develops and becomes more and more individualized. Over time the animals adapt and become rooted to their place. In the intestines of the ruminants a special ferment pattern develops that is adapted to the flora of the place. This shows up in higher health of the animals and in a better performance. For example, early tests on farm organisms have shown that after turning to biodynamics cows use less food protein to build up a pound of milk protein. The animals "make more of their food."

The inner health of the animals radiates back though the manure into the plant world. An organism is born and develops in time. As with any organism, the farm organism needs a strong inner circulation of substances. Depending on the quality of the soil, the farmer has to determine what percentage of the acreage can be used for market crops and what part has to serve the inner circulation through green manure and fodder crops. The condition and the amount of organic matter or carbon substance are decisive factors for the health and the productivity of a farm.

Many other authors have written extensively on the concept of biodynamic farming (see the *Bibliography*).

Free Land

For the farms of tomorrow, land cannot be used as a commodity or a tradable good, like a car or a pair of shoes that are produced, sold, used, resold, and finally used up. After all, land lacks every attribute of a tradable good: it has not been produced, it cannot be produced; it is limited in size to the surface of the earth; and access to it is essential for every human being. The basic economic reality is the relation between the population of a given region and the amount of usable land in that region. No one should possibly have more of the fruits of this earth than what grows on the amount of land that is arrived at when you divide the amount of usable land of a region by the number of people living there—and no one should have less.

The widely held belief that there is such a thing as private ownership of pieces of our living planet is a fiction, a social lie. As any real estate attorney would agree, when one holds title to a piece of land, one actually holds a bundle of rights: the right to use a piece of land exclusively or in cooperation with others unlimited in time, and the right to hand these rights of land use on to successors. If you buy and sell land, you are actually buying and selling rights of use, not a commodity, which land realistically cannot be.

In former times it was the right of certain families to hold designated public offices. That right was considered inheritable. The kings of France, with their assumed right to raise taxes, assigned certain families to collect vast sums of money to finance the court's excessive lifestyle. So the right for taxation went into private hands and was executed there for the profit of these individuals. Millions of people died of starvation in rural France by over taxation. Another odious practice of trading rights is to achieve the right to hold a certain office—ambassador, consul, and so forth—by contributing heavily to the campaign funds of political parties.

The right to draw revenues from spiritual production—the copyright—has been limited to most countries to a period of thirty or fifty years. So there is a growing feeling that rights should not be traded limitlessly. In the future this will apply more and more to land titles as well. To give a common example: a family had farmland in use over a long period. They stopped using the land in, say, 1940 and moved out of the area to live their old years somewhere else. They keep the title to this land. They and their children do not care for this land for fifty years. By holding the title they exclude anyone else from the use of the land. The land bushes in, becomes an impenetrable thicket of no pleasant appearance, and eventually a quasi-forest of very low productivity. Suddenly, and without any help from these title holders, an economic boom reaches the area. Land is needed for development. The title holders sell their title, making an enormous profit that is raised on the labor of those that are in need of the land now.

Hand in hand with the fiction of private property versus parts of living nature, we developed the use of land as collateral, the mortgaging of land. But the history of farm mortgage is the history of farm crises, the depletion of land, of deforestation, or erosion. In its 1938 *Yearbook of Agriculture,* the US Department of Agriculture wrote: "Mortgage may be as injurious to a farm as erosion or a poor cropping system," and "it has often been pointed out that certain tenant agreements result in over-cropping and depletion of the soil but it is not always recognized that this same condition may be caused by the terms of a mortgage on a farm that is operated by its owner. The necessity of meeting payments on a mortgage . . . has caused many farmers to specialize in crops that ultimately reduce soil fertility, to neglect the restoration of humus, and to fail to plant crops that prevent erosion. At the same time the farm house and the other farm buildings as well as the soil deteriorate."

Property taxes place the same kind of negative pressure on farms. Whether the farm has a good year or bad, whether there is drought, or hail, or general crop failure, both mortgage and tax payments come due. They must be paid or the property is forfeited. This again forces farmers into growing market crops that promise a short-term profit. The USDA report of 1938 points that out very clearly: ". . . too great reliance on property taxation in rural communities tends to promote short-sighted land use, which if persisted in, brings about serious deterioration of the land resource."

As the socialist and communist nations of the world have demonstrated unequivocally, state monopoly of land is equally disastrous. Government control destroys individual initiative and with it the land itself. For the farms of tomorrow to succeed, any form of government ownership or even administration of land has to be strictly avoided. Government is unable and unsuited to deal with this task.

The farms of tomorrow must be based on a new approach to land. The land can no longer be used as a collateral for debt; it should no longer be mortgaged. It must be free to serve its original purpose: the basis of the physical existence of humanity. How will this be possible under a Constitution that sees land as a commodity and protects the property rights of the title holders?

The answer is straightforward. The land has to be liberated out of the insight and actions of citizens who recognize the essential need for "free" land. Specifically, local land suitable for agriculture must be gradually protected by land trusts. To do this, every piece of farmland has to be purchased for the last time, and then, out of the free initiative of local people, be placed into forms of trust that will protect it from ever again being mortgaged or sold for the sake of private profit. Non-profit land trusts must then make the land available to qualified people who want to take it into ecologically sound uses. Such arrangements will give the right of land use to individuals or groups, either for the time they are willing or capable of using it, or in a lifelong contract that could even include the right to find one's successors. (See *Appendix B* on land trusts).

The US has a growing land trust movement. These trusts take land into their legal embrace so that it can serve the two basic needs of humanity—ecologically sound farming systems and affordable, non-speculative housing—in a way that excludes profiteering. It is neither necessary, nor likely, nor desirable for legislatures to change constitutional or other legal rights relating to property. The soundest way into change for the better, is for property owners, out of an understanding of the necessity for free agricultural land, to gradually and freely donate or sell their land into such trusts. Landowners themselves could form such trusts, or groups of citizens could cooperate locally to buy the available land for ecologically sound farming and for affordable housing projects. This is clearly something that cannot be legislated or otherwise imposed in any way upon humanity. Every step of progress will have to arise out of the insight and the free initiative of the people. Our expectation towards legislators and government administrators can only be that they will not hinder such development, and perhaps further it in some ways.

The transfer of more and more land from private ownership into nonprofit land trusts will raise many serious financial questions. For example, if the land cannot be used as collateral by the farmer, if it cannot be mortgaged, new ways of farm financing have to be opened up. These new financing mechanisms will be based more on personal credit. This personal credit will be obtained more easily as more people are responsibly connected to a farm operation. In this case financing can arise from guarantor communities of people who are directly connected to a local farm project, as with many of the community supported farms described in this book.

Another great question arises out of the fact that, by and large, in the U.S. today a farmer's financial safeguards for old age and illness are based on the market value of his farm. Against this market value, farmers in need can borrow money and in the end, when they sell the farm, finance their old age. If a new way of financing these needs cannot be found, a new approach to land property and land use would be difficult. In a community supported farm system the cost of illness and old age of the farmers can be met through all the household incomes of those households that are connected to the farm. They become a part of the operational cost of the farm. This can be done with or without the conventional insurance techniques.

The land question is closely connected with the social welfare question in general. For most people, their only source of financial support after retirement is directly linked to the real estate they own, usually their home or their farm. The questions of financial safeguards against illness and old age go far beyond the farmers. But to heal the many social problems we have now, we must begin with the farmers. They, after all, care for the fundamental source of renewable wealth: the land.

Furthermore we have to take into account how much our whole system of money and financing today depends on our mortgage system. The "miracle" of economic growth with its credit financing not only of investments but also of consumer goods, in the way of "consume today, pay tomorrow" is, besides giving credit on income, mainly based on mortgaging land and its buildings. Consequently, the volume of credit can be expanded insofar as the market prices for land and houses rise. But whenever the real estate market stagnates or declines, we inevitably witness the high indebtedness of the farms leading directly to a massive crisis with thousands of farm foreclosures (1928–35, and 1986 into the 1990s). This is clearly an unstable and inherently dangerous system, which causes tremendous hardships for millions of people. The financial system, especially in regard to mankind's essential agricultural activity, must be restructured. Our present money system has, to a large extent, led to the ruination of our farms, our food quality, and our environment.

Ultimately, we must develop a nationwide system of "free land" for agriculture, as described above. This system of free land that cannot serve as collateral will, obviously, imply a different money system altogether. That is one of the critical challenges of the future.

People

In olden cultures such as the classical Greek and Roman, work on the land was generally done by slaves. In Greece this slave labor was provided by the original inhabitants of the land who were overcome by the Dorians invading the peninsula from the North. In Rome the labor was done by various subjugated peoples. The dependency of the peasant population and their slave-like status in society, continued far beyond the Middle Ages. In Russia, certainly, it continued into the nineteenth century, and the same can be said of sharecroppers in America. More and more through various reforms, dependent peasants became independent farmers. But these legally independent farmers soon came into the dependency of financial institutions. The very act of acquiring a farm through mortgaging led directly into this new dependency. The farm laborers became hired personnel. They had to sell their labor for money to make a livelihood, and they still do today.

The farms of the eighteenth and nineteenth centuries in North America were carried usually by large families and the help of hired—often migrant—labor. With the concept of the family farm we connect still certain emotional, moral, and social qualities. So saving the family farm is often a declared aim of politicians and agricultural writers. They cling to this model, even as it declines in the face of more and more agribusiness—capital corporations that manage plant and animal production through hired management and labor. The defenders of the family farm seem often to forget that what they idealize was in the past mostly based on large families, often three generations, working together on one farm, substituted even by hired labor. Unmarried sisters and brothers often stuck to the farm and were welcome and cheap labor. This clan-like family structure is no longer prevalent. The family farm of today means usually one couple managing a far too big land mass and far too big animal operation by themselves and at the same time raising a family. Usually the women in these operations are totally overburdened. Hired labor, no longer available nor affordable, is replaced by machinery. With machinery comes debt and mortgages; the freedom to do what is right on the farm vanishes. So the family farm can no longer be the ideal.

Experts estimate that 25 percent of the usable land in the US will be managed by agribusiness by the year 2000. The alternative to this trend must be a new association and cooperation of families and individuals, excluding the use of hired labor. The typical attitude in farming is a complete devotion to the cause, to the work. If you are not totally dedicated to the animals and plants, to the whole farm organism—without any reservation—the animals and plants will not thrive properly. The hired hand with his wage contract cannot develop the necessary devotion.

In some community supported farm systems, member households carry a certain part of the budget out of various sources of household income. The life needs of those who principally work on the farm, and who have no outside income, are supported by the household incomes of the other farm members. That means that the farmer's family is not driven to make a profit with which it can support itself, nor is the farmer hired for a wage to work for the others. He has been brought into a position where he can donate his labor out of his spiritual intentions for the well being of the farm organism and the fellow members of the farm community. Beyond that every other member family of this community supported farm can come occasionally, or regularly, to donate some work for the benefit of the farm and the community. Beyond this, donating work to the farm benefits the community member by giving him or her the experience of working in nature and the attendant opportunity to deepen an understanding of nature and of him or herself. In such a situation people can experience the possibility of a new relationship to labor. And in considering this possibility, we can gain a sense of how humanity develops in its relationship with labor: from slave labor, to hired labor, to donated labor.

The cooperation of free individuals on free land to ensure the basic needs of all has another ideal background. As stated earlier, land use is a basic necessity for every individual. Without the right to use land, no individual has security. Since the amount of usable land in an area is constant, and the size of the population at a given moment is also known, we can exactly determine at any given time how much usable land is available per person in a given area (in Central Europe about half an acre per person, in America about three acres per person). This ideal/real relationship between land and people is a basic feature of our economy.

The right of the use of a certain amount of land for every individual can be supplemented by the responsibility of every adult for a given area. This right of land use and this responsibility can be shared. People can associate to share their rights and their responsibilities, and so create a true farm organism wherein they will share the care and the cost of the operation. This can be one theoretical basis for community supported farms now, or on the farms of tomorrow, where any given community supports itself through farming.

Essay 3

Ten Steps Towards the Farm of Tomorrow

Many authors have written extensively on the concept of biodynamic farming (see Bibliography). Here, in brief, we outline ten basic steps drawn from this concept that underlie the farm of tomorrow.

The first step for farmers and gardeners should be to remain in the realm of the living with all measures and applications. This excludes most all mineral and synthetic substances for use on plants, the soil, or animals, including mineral fertilizers, synthetic pesticides, herbicides and fungicides, and mineral supplements in animal feeds. Life processes can only be generated out of substances already filled with life. Soil fertilization should always mean enlivening the soil with living substances: manure, compost, and all sorts of green manures. Minerally fertilized soils lose their life capacities (which can be easily measured in soil bacteria per cubic inch), and their capacity to store water and air. As a consequence of this, more and more energy must be used for mechanical soil preparation and for water soluble mineral nutrients.

An exception to this rule comes in the use of lime, such as ground limestone or Dolomite, on acid soils. But even here experience has shown the lime substances that are closer to the life process, such as ground seashells and calcified algae, are more effective; smaller amounts can be used than in the case of ground limestone. But if you do use ground limestone, it can be better and more effective to lead this mineral substance through the composting process by adding a certain amount of it to the organic composting material.

In animal husbandry, mineral supplements to the feed can be almost totally avoided if we provide the animals with a wide spectrum of plants. For the ruminants these should include the availability of perennial plants like bushes and trees with their leaves. Such perennials,

Reprinted from the original *Farms Of Tomorrow*

together with herbal plants, commonly have a higher concentration of mineral substances in the finest preparation. Generally herbal plants of the families labiatae, compoitae and umbelliferae, should play a larger role as feed supplements. One exception to this rule in animal feeds is salt which might be necessary as a supplement out of the mineral world. Pigs can be kept free of mineral feed supplement if they are kept in contact with living soil with its rich antibiotic fungus flora.

Experience has shown that there is a great advantage of always staying with measures in the realm of the living. In using on field and farm only substances that are derived out of life processes one can hardly ever overdose or create any harmful effects. To overdo the application of compost in the field can be wasteful, never harmful. The same is true for plant-derived sprays which may be used to regulate insect attacks or fungus diseases. By contrast, just think what immense poisonous effects appear from the wrong application of synthetic herbicides, fungicides and insecticides. Mark the rule: in all farm or garden measures and applications, stay in the realm of the living.

The second step is to arrive at the manure that is necessary for healthy plant growth by keeping on the farm a sufficient number of animals in the right harmonious combination of species. One can have a balanced farm organism keeping mainly cattle, because of the harmonious characteristics of its excrement that relate in a special way to the fertility of the soil. However, it is better to have a mixed population of different animals that include birds (chickens) and pigs, eventually horses and sheep or goats. The character of each animal manure is different. While cow manure is very balanced with an emphasis on calcium, pig manure has a potash character, and bird manure has the character of phosphorous. These specialties of the different manures can be used in special plant cultivation. Generally, it can be recommended that the different manures be mixed and processed together.

The quality of animal manures depends largely on their diet. Only a mixed diet that is specific for any species gives us a manure quality that serves the soil in a way that will produce the best plants. The time of passage of feed through a cow can vary greatly. However, protein-rich feeds and silage reduce the passage time, and lead to a manure that behaves differently in fermentation. These often smelly and more liquid excrements attract flies and decompose differently. If ever you have grazed cattle and sheep in pasture that includes bushland and forest, you can observe the totally different quality of excrement.

A broad and varied diet that is supportive of the various animal species is the basis of a balanced manure and, consequently, healthy plant growth. The right number of animals and the right mixture of animals has to be found by observation of the individual situation of the farm. Roughly, one can expect that one cow (1,100 pounds) needs two acres for year-round forage, and can keep four acres fertile by its manure. Of the mineral substances our domestic animals take in with their feed, 80 percent is recycled through the manure into the farm.

The third step is to feed the herds—all the animals on the farm—from feeds that are grown on the farm organism itself. This step allows the farm to build its strength and individual character as an organism. The exchange of the feeds from the farm organism to the herds, and from the herd's manure back to its soil, creates a process of mutual adaptation. This adaptation finds its expression in the microbiotic life in the digestive tract of the animals that gets shaped by the specific flora of the farm (a fermentation pattern), and leads to higher performance and a higher level of health in the animals. One expression of this performance is the capacity of the animals to bring forth new life. The average number of calvings in most industrialized countries is between two and three calves per cow-life; the average in a well-developed farm organism with a well-adapted herd should be six to seven (of an animal that can, ultimately, come to fifteen calvings per life).

The aim and the result of this local interchange between animal and forage is permanent fertility. In keeping the necessary number of animals on the farm we necessarily introduce more fodder crops into our rotation. These fodder crops, such as alfalfa, clover and grasses, provide a great root mass for the upkeep of the organic substance in the soil, and help create a soil based on life, not fertilizers. A great variety of fodder plants should be used to serve the needs of the animals for a balanced diet and to create balanced crop rotation that keeps up the healthy fertility of the soil. Practically no feeds should come from outside the farm organism.

The fourth step is to aim for a great diversity of plants on the farm in combination with, and as part of the crop rotation. Fertility and productivity in nature arise out of diversity, not out of specialization or monoculture. The author counted up to seventy plant species in use on one of the biodynamic farms he served, including six grains, eight leguminous fodder plants, twelve grasses, numerous brassicas, vegetables, and herbs.

By contrast, a modern dairy farm in New England cultivates, beyond its hayfields, usually only one species: corn.

A well-managed pasture should have, besides grasses and clovers, deep-rooting plants like alfalfa and herbs. They mobilize in the soil mineral substances from lower levels and make them available to other crops. The management of such pasture should be so that the root growth of the plants is not inhibited by overgrazing late in the year. Plant rotation in the fields can be looked at as a diverse plant community spread out over time. Here again, deep-rooting plants that leave behind masses of organic material should alternate with shallow-rooting plants.

For the feeding of mankind, the total production of our farms counts more than a high level of yields in special crops. Experience with many organic and biodynamic farms has proved that their total output of food, measured in calories (Joule) is generally higher than the output of conventional farms, where the yields in their single market crops can be substantially higher. The reason for this is in the diversity of crops on biodynamic farms and the productivity that results out of this diversity. Plant species support and complement each other more than they hinder each other.

The fifth step is to recognize that the circulation of carbon, or organic substance, throughout the soil, the plants, and the air is the basis of permanent fertility. This circulation expresses itself in the creation and breakdown of humus substance in the soil. Consequently, great care must be given to carbon circulation, the proper break down and build up of humus. Experts estimate that through the last 150 years, 50 percent of the humus substance of our fields has been lost—either used up, or washed away through erosion. The life processes of the soil microorganisms, earthworms, springtails and so forth depend on the functioning of this living humus layer of the earth. And humus is built up by animal manures and plant residues (root masses and green manures). To keep soils from being exploited, turnover of an estimated 30,000 pounds of carbon matter per acre per year is necessary and sufficient. To achieve this we have to find out for each farm what amount of the crops can be sold and what amount has to be green manures that will go into the humus circulation. The acreage of the farm that can carry a market crop will necessarily be smaller on a poor soil than on a fertile soil, versus the acreage of crops grown for green manure.

An essential tool in keeping up the humus level is the right or sweet fermentation of animal manure, and the composting of plant matter. To guide these fermentation processes in the right way one can use herbal composts that are implanted in manure and compost piles. These special herbal composts can be prepared by the farmer, or purchased from Josephine Porter Institute For Applied Biodynamics, Inc. (see *Appendix F—Resources*). Another great factor in the care for the humus layer is cover cropping and adequate tillage. If we have nutrient deficiencies on an organic or biodynamic farm, we do not work on the nutrients directly by applying them, but instead we work on the humus level and the humus quality.

The sixth step is to strengthen silica circulation in the soil by encouraging microbiotic processes. As we deal with the carbon cycle and with the life forces and the humidity in the soil, we also need to pay attention to the circulation of silica, which is linked with the light and warmth forces in the processes of nature. The Russian scientist V.T. Vernadski declared "No doubt is possible about the fact that no living substance, no organism, can exist without silicium."

The ashes of many trees and of the high grasses contain large amounts of silica. In most other plant species, silica plays an important role in the roots. In spite of this, silica can be called the forgotten element in agriculture. Among the fifteen elements that are most frequently named as essential for healthy plant growth (carbon, oxygen, hydrogen, nitrogen, sulfur, phosphorous, potassium, calcium, magnesium, iron, boron, molybdenum, manganum, copper, and zinc), silicum is not mentioned.

One reason for this is that silica is not solvable in water. It enters the life process of plants only through the living activity of soil microorganisms. If enough silica is taken in by the plant, more happens in the plant through the influence of the sun because silica transports light and warmth. Both directly and indirectly, the sun forces and other cosmic influences in their different qualities are transferred to the plant through the silica. The parts of the plant that contain much silica grow twice as fast as the plant organs that contain more calcium (Voronkov, *Silicium und Leben,* Berlin, 1975). One of the problems of one-sided fertilization of plants is that it changes their mineral composition drastically by reducing their silica content. As chemical analysis of plant ash has shown, between 1840 and 1940, the silica content in plants has been reduced drastically, and we can assume that in most commercial agriculture the downward trend illustrated in the following table has continued into the 1990s.

Silica Content of Plant Ash, 1840 vs 1940

	Wheat Straw	Rye Straw	Barley Straw	Oat Straw
1840	81%	82%	71%	79%
1940	63%	48.2%	52%	46.5%

(Duftos, 1840 and Menzel Lengerke, 1940)

In consequence of this reduction in silica, the straw looses its resistance and strength, becoming prone to fungus infection. If the silica content has been reduced in the straw, then one can also assume that a similar reduction has taken place in the grains eaten by livestock and human beings.

High intake of mineral salts, caused by the use of petrochemical-based fertilizers, in all probability in turn causes the reduction in silica content. The reduction in silica content parallels a heightening of potash content. Modern agriculture tends to push the growth of plants with the use of nitrogen-rich fertilizers, but this changes the inherent balance of substances within plants and frequently makes them prone to pests and diseases of all sorts. The crystalline substance silica is essential for healthy plant structure, and also for human and animal health. But silica can only be absorbed gradually into the plant through the aid of the microbial life in the soil. For that reason, rather than feeding nitrogen-rich fertilizers to the plants themselves, the farms of tomorrow must gradually encourage the life in the soil with a broad and balanced supply of organic matter. As the microbial life of the soil thrives, so the plants that arise from the soil will be able to take in a balanced quantity of silica and so thrive.

The use of finely diluted and specially prepared crystal silica promotes balanced growth processes in green plants.

Step seven is to create harmonious balanced conditions in soil, plants, animals, and landscape as the necessary basis of productivity. The health of plant and animal organisms is a function of this harmony.

Out of the six previously mentioned steps we can begin to see that productivity in nature arises out of a harmonious, balanced situation on all levels. This situation we have to achieve in the landscape, in the number and mix of domestic animals on the farm, in the building up and the breaking down of humus substance, and in the metabolic processes in the animals and the plants.

There is a polarity of warmth and light processes to water and earth processes, and likewise a polarity of warmth and light substances to water and earth substances. In testing a wide variety of plant material, the great American soil specialist Firman Bear found that every species has a specific relation between these substances, and that this establishes the pattern of the plant.[1] The plant is not healthy and nutritious because it has an abundance or deficiency of certain substances, but rather because it has all of its substances in harmonious proportions and order or not. There has to be kept a certain balance between the substances which build base and those which build acid. The balance is typical for each plant, and it can easily be destroyed by unbalanced mineral fertilizing, making the plants prone to insect pests and fungus diseases.

From the landscape we have to work toward a balance between forest, field, pasture and wetland down to the inner structure of the plant. We must work and produce out of harmonious balanced situations.

Eighth, the destroyed natural environment must be restored. In particular, for this restoration to occur we must pay attention to two elements: hedgerows and wetlands.

The hedgerow is the natural habitat of many valuable birds and insects, a natural barrier against wind erosion, and also an invaluable aid to maintaining humidity in the soil. Birds and other creatures which find a home in hedgerows feed upon harmful insects and so help to maintain the natural balance. Valuable insects also live in hedgerows and on the fallow land beside the fields. Ant colonies, for example, often establish themselves in hedgerows. The ants may range up to 200 feet from their nests, and are a reliable control for the larvae of the potato beetle. Hedgerows also provide a source of wild fruit for man and animals. By nibbling on the hedgerow, cattle diversify their diet and so maintain optimum health. Hedgerows should not be established as monstrous lines of trees, but rather should consist of a great variety of bushes and trees so they will attract a great variety of birds and insects. Many blossom- and fruit-bearing varieties should be incorporated into the hedgerows, such as quince, filbert, elderberry, and fruit bearing roses. More and more people will in the future want to supplement their diet with wild fruit.

Another healthy factor in the landscape of a farm organism can be ponds and wetlands. They enhance the production of dew, which supplies water to the plants during dry periods. They are also the habitat of amphibians, and of many insects and birds which are beneficial for the environment. The moist environment of the wetlands attracts microorganisms and insects, thereby distracting them from the croplands.

Consequently, with ponds and wetlands we can again expect and observe a regulating effect in the whole ecology of the farm.

Step nine is to implement biological weed and insect pest control. As described in step eight, a sound diversified ecological system that contains forest, field, permanent pasture, hedgerows and wetlands, with numerous domestic and wild animals, is the basis of biological weed and pest control. A second basis is to develop healthy humus content in the topsoil. The antibiotic fungus flora of a developed humus has a cleansing, detoxifying influence on unprocessed, or raw organic substances. To keep certain insects under control it is nevertheless important to bring very little unprocessed organic substance, such as raw manure, into the soil. The most effective procedure against weeds and pests is a diversified plant rotation in the fields. Finally, over the last sixty-five years many steps have been taken to isolate out of plant and animal organs substances that help to reduce the pressure of weeds and insect pests. Those substances can be used to good effect, without ever having to resort to chemical poisons.

Step ten is to reestablish a rhythmical order in animal husbandry and field care that is connected to the rhythms of the earth and its cosmic environment of the sun, the moon, and the other planets. This is an essential part of farming in the future. Life is rhythm. All life processes develop in rhythms that are related to cosmic rhythms. The moon rhythm is obvious in the gestation periods of man and the domestic animals. The sun force in summer and in winter constitutes different influences on the lower animals and on plant life. Research into these rhythms is just at its beginning, but numerous planetary influences on plant and animal life have been found and documented over the last fifty years. Much of that research can be found in the publications of Maria Thun, especially in her annual calendar *Working With The Stars,* in the *Kimberton Hills Agricultural Calendar,* and in Lilli Kilisko's *Agriculture of Tomorrow.*

Following the basic principles of these ten points will yield ever greater individualization of every farm. By restricting, and in some cases eliminating altogether, the import of substances foreign to it, the individual farm organism protects its inner harmony and rhythm. The steady inner rotation of substances and seeds, and great diversification, lead to a healthy adaptation of plants and animals to the specific micro-environment of the farm organism, and to high productivity. As recently as 100 years ago, the tiny country of Switzerland produced over a hundred varieties of wheat—each variety specifically adapted to the micro-environment of one of its valleys. Today, by contrast, only about a dozen varieties of wheat are grown worldwide, and those varieties need to be replaced every ten years or so because they have become susceptible to pests and diseases. Such a generic, mono-cultural approach not only narrows the diet of humans and their domestic animals, it also makes the worldwide food supply extremely vulnerable.

The economic goal of the biodynamic farm is to have rising quality, and diversification of products which are adapted to the need of the local population. Meanwhile, the input of foreign substances and energies goes toward zero. Such an approach has been proved to be possible in hundreds of biodynamic farms all over the world and is the only real, undiminishing source of support for mankind. While not all of the farms described in this book employ the biodynamic approach, elements of biodynamics are to be found in every one of them.

1. Firman Bear, *Soil and Fertilizers: Cation and Anion Relationship and Their Bearing on Crop Quality.* New Brunswick, NJ: Rutgers University, Department of Soils.

Essay 4

Three Basic Rules

Many things written in this book would appear illusionary and unrealistic if there were not already examples of farms where, in one way or another, all of the ideas mentioned have been realized for some time. Immense social, economic, and legal obstacles lie in the way of structuring our farms in new ways, and restructuring generally and specifically our relation to this earth. However, those who examine the situation will recognize that the crisis of our food includes deteriorating quality as well as insufficient availability for many millions of humans. Realizing this crisis, and the general degeneration of our natural environment, and seeing also the education crisis and the moral insanity of our time, we have not much choice. We must reestablish our relationship to the basic sources of our livelihood; we can do this best through helping to create the farms of tomorrow and bringing ourselves into a new relationship to them.

The farms of tomorrow will not be just family farms. They will be so complex and diversified that they will require the cooperation of many different, unrelated, free people. Such farms will need the cooperation of several households as well as the open support of households whose members are not actively farming but who share the responsibility, the costs, and the produce with the active farmers.

Three basic rules for making such cooperation possible were once given to author Trauger Groh by Wilhelm Ernst Barkhoff, a lawyer who had a deep interest in and personal experience with these questions. Trauger relates them here in his own words and understanding, because in many years of farming he has found them to be true and helpful.

The first rule is, do not work too many hours. Farming is labor, craft, and art. The art arises out of a deeper understanding of nature based on thorough, ongoing observation, reflection, and meditation on all surrounding natural phenomena and processes. If the active farmer is working too long hours, he lacks the leisure for this observation, reflection, and meditation. He loses his art, handing it over to extension services and agricultural schools. They themselves transform the art of farming, which is the true agriculture, into materialistic natural science; and they transform the inherent, developed economics of the farm organism into a merely profit-oriented exploitive agronomy of plant and animal production. Finally the farmer is no more creative himself, he is just applying their recipes. In the end he loses a lot of his craft, too, depending less and less on farm skills and more and more on the supply of sophisticated machinery. He often becomes a badly paid, highly indebted laborer with, hopefully, high technical skills.

The second rule is, buy for the farm as little as possible from the outside world. The less you purchase in the way of sophisticated tools, machinery and farm buildings, the more you are financially independent and free to work with and out of nature. Use the help and the skills of all friends who are related to the farm. The secret of successful Amish economic life lies in relying on human beings to work the farm and not buying too many things. If you work in cooperation with other people—and that is what the farm of the future is about—you will find that purchasing costly tools, machines, and buildings from outside strains the community; it raises social questions. The initiative to purchase these items usually comes from an individual. The consequences of the purchase have to be carried by the group. This raises problems.

The third rule is, take all the initiative for your actions on the farm out of the realm of the spirit, not out of the realm of money. What does this mean? The more we penetrate the spheres of nature, the more we become aware that what surrounds us in it is of an overwhelming wisdom. What we call scientifically an ecosystem is penetrated by wisdom so that all parts serve the whole in the most economic way. In awe we stand before the higher "intellect" of a bee-hive, or an ant colony. The cooperation of microorganisms, earthworms, springtails, and others in the soil in breaking down plant material and in building humus particles, is something deeply rational and wise which cannot be copied or synthesized by man. This we can describe as the spirit that is spread out in nature.

This spirit organizes nature with the highest economy. For example, the ratio between consumed substance and achieved effect in a bird that migrates from the Arctic to the Antarctic is of a scale that men cannot achieve technically. The comparison between bird flight and an airplane demonstrates this clearly. The calcium structure of a hip bone makes it possible to carry a maximum of weight with a minimum of substance, a structure that again surpasses our technical capabilities.

The more we understand and follow this "wisdom" in nature, this outspread "spirit," the more rationally and therefore economically we can organize the farms of tomorrow. The profit motivation, applied to nature, has led to vast depletion of soil and dangerous exploitation of animal and plant material. If we follow the spirit in nature, we put into our service both the rationale and the economy of nature. This, ultimately, is the basis of the life of humanity.

What practical steps can we take now to make the farms of tomorrow possible? We have to bring together, piece by piece, and lot by lot, agricultural land, and then see that it becomes free of speculation and mortgages. We have to establish and support training programs closely connected to farms that are under ecologically sound management, to train and educate the future farmers. And in connection with such farms, we have to establish research programs that support all these aims.

Thousands of men and women in the US own farmland without using it as collateral and without needing to give it over to development. To them we have to appeal, and to those who can give surplus capital to acquire such land. Land given or purchased under this concept must be legally secured so that it can only be used in ecologically sound ways, and so that it cannot be given into development for personal profit. Land trusts have already proved themselves to be one suitable form for this purpose, and a network of land trusts already exists as a supportive model (see *Appendix B*). To prepare the necessary organizations and to start a campaign toward these ends is an urgent need of today.

These aims are only possible if we create, on well-run farms, training programs that enable young people to acquire the necessary understanding for ecological farming, together with the necessary skills. This cannot be done in the form of apprenticeships only. It needs, besides training through work, classroom studies and nature studies with qualified persons. The cost for such training cannot be carried by the farms.

Finally, we will have to establish a new way of agricultural research. The task of such research can be defined as studying the social and economic conditions of ecologically sound farming, and also of farming concepts themselves, such as organic and biodynamic agriculture. According to this definition, the new farms described in this book already constitute, in every sense but the formal, an essential research program.

This book was written not to show final results, but rather to point out some approaches leading toward the farms of tomorrow, farms that are needed today.

Economic, Legal, and Spiritual Questions Facing Community Farms

Entering into the adventure of a community farm—a farm that is both a community effort and a community responsibility—leads us into manifold new social relationships. There are, for example, the relationships among the active farmers, among the active farmers and the farm members (or shareholders), and among all people who benefit from the farm and share in some way the risk, or cost, of the farm. There is also the growing relationship among community farms, one to another, between the farm operation and the land owners, and among these farms and farm associations and the larger economic and social environment. All together, it is a sizable network of relationships of a threefold nature: economic, legal, and spiritual. Into all of this, we have to try to bring some consciousness, some light.

Let us start with the economic dimension. As Gary Lamb has pointed out in his writings and speeches, the community farm movement embodies elements of a new associative economy that is fundamentally different from the ruling market economy.[1] Associative economy means that all participants in the economic process try to listen to the needs of all other partners in the process. The active farmers listen to the needs of the member families. The member families listen to the needs of the farmers. One community farm associated with the other community farms in a bioregion listens to the needs of the others. On this basis they proceed.

The market economy is driven by the self-interest of every participant. That means: I produce at the lowest possible cost, and try to sell at the highest possible price, independent of the true needs of my customers. And I try to buy the supplies needed for my farm at the lowest price available, independent of the true need of the producers. The ideology of this market economy claims that by this attitude and approach the greatest welfare of the largest number of people is achieved.

In an associative economy, we associate with our partners—active farmers among themselves, active farmers with all the member households, farm communities with other farm communities. The prevailing attitude is a striving to learn the real needs of our partners, and the ways we can best meet them. That means we do not make our self-interest the driving force of our economic behavior, but rather we take from the needs of our partners the motivation of our economic actions. We believe and trust that this will lead to the greatest welfare of all involved.

Associative economy is a complete contradiction of the ideology of the market economy as it is sold today as the all-healing method for the woes of humanity. By developing associative practices in the community farm movement we introduce a new, important and necessary element to the whole economy.

For example, a farm or a group of community farms may be located in an area such as the Northeast of the United States, where grain growing is difficult (often a grain harvester is no longer available) and sufficient grain quantities and qualities can rarely be achieved. The farms need grain for their members, and to feed their livestock, and so they turn to an organic farm in an area such as the Midwest that grows high-quality grain. The market-economy approach would be to search for and identify the farm that can supply the grain for the lowest price. But under an associative economy, one possible approach would be to formally associate with the grain farmers in a way that allows the Northeast farm to learn and understand the true needs of the Midwest farm to cover the cost per acre of the grain they grow, then to meet those needs and share in the risk of crop failure or the benefit of an abundant crop yield.

Experienced community farmers all know that we have to adapt our planning to the needs of our farm members and their households. This has become clear over the last ten years. We can educate our members about the new and unusual varieties of foods that we think are good to grow upon our land, but we cannot force them to eat these foods. We can expand certain crops, and we can shrink them. But we always have to put the human being and his or her needs at the heart of our economy. As we have learned from experience, it just cannot work any other way for a community farm. This attitude—putting people at the heart of one's efforts—is the basis of associative economy.

To date, this wisdom is not applied in the mainstream global economy, and its harsh consequences are coming plainly into view. In 1996 an association of 2,300 French business executives made this point in a landmark paper: "For 20 years, companies have become profitable at a cost to society. . . .We are convinced that non-regulated capitalism will explode as communism exploded, unless we seize the opportunity to put man at the heart of our society. . . .What is needed is a fundamental rethinking about the way work and society are organized."[2] As the French executives have seen and warned about, the mainstream capitalistic economy puts investors at the heart of the society, with scant consideration for human beings. Such an approach has inevitable negative consequences.

Associative economy was first expounded by Rudolf Steiner in a number of public talks and writings in the years from 1917-1919, and has been developed in theory and practice by many others in the following decades.[3] It shows that we must put the needs of our fellow human beings into the center of our activity: identifying needs and covering those needs with the least effort (the least input of substance, energy, and labor). That is true economy, and only that is economy.

Where are the limits of this associative adaptation to the needs of community farm members? The limits are in the agreed spiritual-agricultural concept that is carried by the farm. Foremost the active farmers, but with them the whole farm community, require an ideal or vision to follow, such as a healthy environment, with high soil fertility that is based on balanced crop rotation and sufficient livestock.

We cannot, for example, enlarge the amount of acreage for root crops at the cost of other crops without bringing imbalances to the soil such as depletion of organic substance (humus), depletion of potassium in the soil, and other serious consequences.

Here we bridge from the economic dimension of the community farm, to the spiritual dimension. What does this entail? The whole planning of the farm organism—landscaping, crop rotation, animal husbandry, seed conservation, and other aspects of the planning process—should proceed in service of the higher ideal of the farm organism. This includes the establishment of an annual budget that serves these tasks, expresses the underlying ideal, and shows its cost.

In this ideal or visionary realm, the active farmers are the custodians. Under the precepts of associative economy, they are autonomous: nobody can give them any order or any advice if it is not asked for. While autonomous in the development and application of their ideal, they must be public in its description. The active farmers have among themselves to shape and work on their ideal free of the initial limits that may appear to arise from economic and legal considerations. Then they have to learn to describe this ideal to the public (the supporting members, or shareholders of the farm), and to describe every step, every year that leads toward the ideal via the budget, knowing that the ideal, by its very nature, never can be reached completely. Without this shaping of the ideal, and its careful description, the community cannot and will not support efforts to attain the ideal.

Through visiting many farms in America and abroad I have found two extremes. Farmers who have an ideal that does not come into praxis, and farmers who are overwhelmed by daily chores—a surplus of praxis and no ideal or vision that they are working toward. Spiritual life or ideals that are not practical are as useless as a praxis that is not penetrated by spirit.

That leaves the third dimension of the farm community (and, really, every community): the legal-social field, the realm of human relations. This realm must be considered whenever two or more people are involved in an undertaking. By way of example, consider the fictional tale of Robinson Crusoe. At the start of his shipwrecked adventure on a tropical island, Crusoe has no rights life because he has no partner. Yet the moment that Friday arrives on the scene—the moment there are two people working together in relationship—the rights life begins. Likewise in all human relationships, even in marriage, rules and social contracts are to be set.

The complex network of relationships that a community farm represents brings into being many legal and social relationships, both written and unwritten. We leave out the normal legal relationship that everyone has to the outer world, the larger society—to the town, to the bank, to the tax collectors, and others—and focus on the inner social and legal relationships of the community farm.

There is the relationship among the active farmers, even if it is as basic as a married couple running the farm, and there is the relationship to the legal owners of the farmland. Different questions will arise, and it is important to distinguish their character. Are they spiritual, economic, or legal questions?

Spiritual questions are those that arise out of our perception of nature, out of our understanding of the farm organism, out of the concepts that arise from these perceptions, these understandings. Such questions are linked to our individual talents, our learning, our education. They involve the ways we apply our concepts to the running of the farm. Some typical spiritual questions are as follows: How do we create a balanced rotation of crops adequate to our soils? How many and which varieties of animals do we need to keep up the fertility of the soils? Do we need to plant trees and wind-breaking hedges? How shall we compost organic substances and animal manure? How shall we till the soils? What tools and machines do we need for these purposes? What buildings? All these questions arise out of our individual insight. They are spiritual questions. The utmost freedom has to reign in this field.

The economic field follows our answers to these spiritual questions, and makes real on earth the preconditions of our spiritual intentions, and their consequences. In economics we address the question, how can we realize our ideals in the world most effectively with the least input of labor and material?

If out of conceptual or spiritual reasons we need a new barn, our economic considerations are how to erect it in the most economical way. The best ways to meet the needs of the farmer, the farm organism, and the members of the community, have to be found here.

What are the rights questions in the farm community? If we have allocated in the annual budget a certain sum for the personal needs of the active farmers, not to pay them such a sum for their labor, but rather to support them and their families, the question arises of how to divide this sum. The fair sharing of the financial result of an operation is a rights question in which all participants have an equal say.

The total sum that is allocated for personal use in an operation, whether it is shared profit or salary, is the financial result of the operation which was created by all participants; thus, its division is a legal act.

If we consider the question of how the farm community shares the cost of the operation, we may consider whether equal cost shares per participant is the best policy, or whether the income situation of the members (shareholders) should be taken into consideration in setting their individual cost to support the farm. Such a question is in the sphere of rights. Here both the opinion, and perhaps the vote, of every participant is relevant. In a true associative relationship, the decisions should not be dictated by just one group, such as the active farmers. Here one can put diverging opinions to a vote if consensus is not reached.

If the farmers need a piece of heavy equipment to realize their ideas, such as a combined harvester, and the financial means in the budget are limited, a decision in the rights sphere has to be made. But never should the initiative come from the economic sphere (for instance, could we make more money with that equipment?). Only the decisions which will implement the initiative which has arisen from the spiritual questions should be based on economics.

Decisions such as setting prices if the farm's products are to be sold outside the community, and the basis upon which products are to be exchanged with other farms (how many eggs should be given for a hundredweight of wheat?) are widely in our economy based upon the market power of the participants. They are truly rights decisions; rights decisions with clear economic consequences that have to be considered.

The decision about which day of the week should be the pick-up day when shareholders come to receive their share of the harvest, is in this rights sphere also. Here the agreement of equal partners has to prevail. The rights of all who are involved must be considered and taken into account.

The decision of whether to buy an agricultural machine, or to buy a good bull for the herd or to spend more money on the livelihood of the active farmers is finally a rights question, based on the understanding of the economic consequences.

Who can justly share in making these decisions? Only those in the community who fully share the financial risk of these decisions down to their personal assets. In too many cases today, this will be only the active farmers. They will be the ones taking the burden of risks to provide food for the community. But it can be and should be more people who are truly interested in the development and well-being of the farm. Our economy, our society, today suffer from the reality that far-reaching decisions, affecting many others, are made by people who share no risk, or only a limited risk.

In any community or society—considering the whole of the economic, legal and spiritual spheres—we can say, so far as farming is concerned: in the planning of the farm, landscaping, crop rotation, breeding, composting, and so forth, down to the planning of the daily work, we are in the spiritual realm. Here should reign freedom and individual perception, even healthy egotism. We have to train ourselves to do this spiritual work freely, without at this point being limited by potential economic and legal considerations. Those concerns must be taken up secondarily, otherwise they block the free flow of spiritual impulses.

In the economic realm we should identify needs, and then work out ways to meet them. The mood should be altruistic, embodying recognition of the true needs of others. And in the rights field we should try as equals to find just solutions.

If we treat these three spheres as their nature demands, we introduce a wholesome, new structure into our community. This structure is necessary because while we surely enter or create a community bringing in our ideals, at the same time we also bring in our lower nature, containing unhealthy ambition, jealousy, greed, carelessness, and so forth. These elements of our nature can easily undermine or wreck the community if we do not have a clear, accepted structure. This threefold structure can be adapted likewise to other institutions, such as schools.[4]

Some may object that most community supported farms are too small to allow any such structuring. To this we can say: we have more and more sizable farm-based communities that are struggling to establish a right social structure. They need not adopt a structure like the one outlined in this chapter, but they might get some ideas out of it. Even for the smallest place, where only three or four people meet regularly to discuss matters of the farm and the farm community, it might be fruitful to consider at every juncture—Where are we with this question? Is this a spiritual question, or a question of rights, or of economics? Every decision we make will have an impact on all three spheres.

For example a farmer may decide to paint his barn stark red. This has a spiritual side. Does this red fit into this landscape, does it go with the colors of the other building, and so forth. The decision will have a social side: his wife and children may hate red, the neighbors may be offended. It has an economic side: the red color costs money, perhaps even more than another color. We have to look at these three spheres separately, not neglecting one, and then come to a decision that is carried by all.

If one works with a group and has regular weekly meetings, it might be helpful to train oneself to the right attitude by deciding, for example: we will speak the first week about the spiritual aspect of an issue, the second week about the economic aspect, and the third week about the rights aspect, and make binding decisions at that time.

But beware: this decision-making group ought not to be allowed initiative of its own. Rather, it decides only about the options that come out of the spiritual planning group after considering the economic and legal aspects. Here the applicable principle of our threefold structure in agriculture is: take the motivation for your actions always out of spiritual considerations; then address the legal and economic aspects.

If we consider what we have said above for institutions such as farm communities or schools, we can get a feel for how beneficial it would be for society if we would set the spiritual life—education, art, religion, even the judicature—free and apart from the influence of the state and political decisions. How beneficial it would be to confine the state to police, defense, and legislation, so far as it does not legislate spiritual or economic matters. Likewise, how beneficial to confine the economic sector to the production and distribution of goods, taking away from it labor, capital, and real estate that, by their very nature, are not and should not be made marketable commodities.

Labor should not be considered a marketable commodity, but rather it should be done out of an insight into its necessity. In farming this is obvious. You can properly care for a farm organism, even for a single calf, only out of interest in it, not for financial remuneration. Capital should not be considered a marketable commodity because it derives out of spiritual activity (inventions) and should not be in the hand of the highest bidder, or the shareholder, but unconditionally in the hand of the qualified user. Land should not be considered a marketable commodity because it is the basis of everyone's living, and should like capital be in the hand of the user as long as the use prevails.

Eventually, we will have to create state-independent bodies (such as boards, or chambers) that—following rules that may be set by a given legislature—regulate the transition of land and capital from the ongoing users to the next qualified users. These boards, or chambers, are part of the free spiritual life, because they have to judge the expertise and qualifications of the next person or group that aspires to the right to use the land, or capital, in question.

It is widely understood in political science that we basically have three spheres of social life: the cultural-educational life, the political life, and the economic life. What has not been seen clearly is that these spheres require different attitudes, and have to be separated more clearly. As once in the ancient theocracies the spiritual-religious life overwhelmed both the rights life and the economic life, so today an economic life based primarily on self-interest overwhelms both the rights life and the spiritual-cultural life. This is the prime cause of poverty and need in large parts of the world, even in America.

This essay is not the place to discuss these macro-social questions in detail. We have to confine ourselves to the necessary social structures on the institutional level, and we have to make a key distinction. As necessary and indispensable as a reasonable social structure is, it cannot solve all social problems. On the personal side, our relationships depend upon the attitudes that we carry toward our fellow human beings. Here we can state as an ideal that we have to learn to accept our fellow human beings totally as they are. That does not mean that we cannot disagree, or that we should not debate in the realm of concepts and ideas; this is necessary and right. But the other person has to be accepted in their basic human dignity and finally loved as the person he or she is. This requires an attitude that allows the other person entrance into our mind and soul without objections and without judgment. By way of example, we can argue that nationalism is a negative force in the world today.

That is a fair and healthy debate. But if we say: "you are a fascist," or go to an extreme and say "you are a Nazi," then we fight not the person's ideas, but rather we attack his or her personality, which should be sacrosanct. This is a very rough example, for there are many more subtle and evil attacks on the dignity of our fellow human beings.

Our social troubles and difficulties have to be dealt with from two poles. At the community or institutional pole with a sound structure, and at the individual pole by a new, elevated attitude toward our fellow human beings. Success at the first pole can be cultivated by an understanding of those different spheres of society. Success at the second pole can be reached only on the thorny way of inner development of the individual. Both are needed to bring freedom, justice and healthy altruism to our affairs.

Finally, we have to look at means to handle the likelihood of grave social disharmonies in the community, such as fundamental disagreements among the active farmers, between the farmers and the supporting farm community, and between the farm community and the land owners. While not inevitable, disagreements and personality conflicts are common in every human community.

One approach to handling such disharmonies is, in every community effort, for the participants to agree in writing at the beginning to establish an arbitration board and to agree, likewise in writing, to follow the ruling of this board. This arbitration board should consist of at least three people who are not involved in the operation, and who have the trust of every single participant. This is better than ending up in a public court where the members of the community might face judges who have no understanding of the community's special operation. As concluded by French executives of the Centre des Jeunes Dirigeants d'Enterprise, "the greatest misery in our society is social and spiritual rather than material."

In my lifetime radical new changes of our social structures and behaviors have come about, mainly through farm communities. Today we can see new and innovative structures arising at farms such as: separation of land property and farm operation (land trust), cooperation of independent (rather than employed) people on a single farm operation, engagement of non-farmers in the responsibility and risk of the operation, a distribution based not on market forces but on the needs of the local and regional community. These innovations point into the direction the whole of our economy and society has to go to alleviate and avoid major disasters in the coming millennium.

1. Gary Lamb, *Threefold Review*, No. 11, (Summer/Fall, 1994), and also his remarks in Example 5 in this book.

2. Centre des Jeunes Dirigeants d'Enterprise, "The Company in the 21st Century." *International Herald Tribune,* Oct. 15, 1996.

3. Rudolf Steiner, *Toward Social Renewal, The Social Future,* and other works. Hudson, NY: Anthroposophic Press, 1993.

4. Dieter Brüell, *Waldorf School and Three-fold Structure.* Nearchus, Holland: Lazarus Publishing, 1995 (soon to be published in the US).

Essay 6

Domestic Animals and the Farms of Tomorrow

Long past the time of the hunter-gatherer societies, when mankind turned from a more herd-oriented culture to a culture that took hold of the earth more and more through tillage and planting, flocks of animals were integrated into the newly developing form of agriculture. Domestic animals became indispensable for draft, milk, meat, hides, and wool; foremost, though, they were indispensable for soil fertility.

For these essential purposes, all farms of the past up to the beginning of the twentieth century included a variety of farm animals. Over the last hundred years farmers have more and more disintegrated the farms by separating animal husbandry from plant cultivation. The animals have been moved into huge compounds, separating them from open nature, and separating the species from each other. Plant production is increasingly carried on farms free of animal husbandry.

This trend is not only prevalent in modern conventional farm systems, but also is strongly present in organic farm systems. A recent survey of community supported farms in the Northeast—all of them managed organically or biodynamically—shows mostly seasonal vegetable operations, and hardly mentions farm animals.

Sir Albert Howard, the great pioneer of the organic movement, was convinced that every farm needs farm animals. Was he right? And was Rudolf Steiner right in proclaiming that a healthy farm has to produce all necessary manure through its own farm animals? What reasons stand against having animals on the farm? And if these reasons are good reasons, how can we deal with them?

The separation of domestic animals from farm land has serious consequences. Therefore, the reasons for farming without animals require thorough consideration. If you ask farmers who farm without animals about their reasons for doing so, you get one or more of nine basic, serious answers. I will paraphrase each of them below:

Based on a lecture presented in Minneapolis, November 1, 1996

1. The land question. "I am farming in a suburban setting where many people live who are interested in organically raised produce. There is not enough land available. The land I have I need for vegetable production. For one animal (a cow, or six sheep) I would need a minimum of two to two-and-a-half acres of land for summer grazing and winter feeding. Since a cow does not want to be alone, if we kept cows and some offspring, we would need at least six acres of land. The green crops in the rotation and the good use of the cow's manure would keep their six acres, plus two-and-a-half acres of vegetable land, permanently fertile. But I do not have that much good land."

2. The income question, and the work force question. "Even if we had some livestock, we cannot support our family and the help we need with a few acres of vegetables. After the growing season we are compelled to take outside jobs to make ends meet. This does not allow us to care year round for farm animals. I do not get help, and if I did get help I could not pay for it."

3. The priority question. "In addition to farming, I wish to lead a cultural life—to be able to enrich myself through lectures, sports, dance, seminars, reading, to go to the theater, and to attend concerts. Likewise, I need to go on vacation, and to dedicate time to my spouse and to my children. Feeding and milking every day would not allow that. I do not want to be cut off from my family, and from culture. At least after the growing season, we want to be free for these other parts of life." (Having been tied to herds most of my forty years as a farmer, I know this priority question intimately.)

4 and 5. The questions of vegetarianism and soil fertility, and the question of killing animals. "We reject the use of animals for human consumption. We reject the butchering of animals that we have raised. I do not want to be included in the killing of animals." This goes together with the fifth reason: "We and some of our customers are strictly vegetarian. We need neither meat nor milk. Those of our customers who need these can get them somewhere else. We have been able so far to keep up our soil fertility with the help of vegetable compost, green manure, and mulches. We use only very little manure, and that comes from outside the farm."

6. The hygiene question. "We do not want domestic animals on the farm for reasons of hygiene. The manure gives bad smells, and draws flies. Our neighbors object to us keeping animals. Animals are unhygienic."

7. The natural animal population question. "We cannot accept the argument that domestic animals are needed to have the animal element in the farm organism. We have deer here nibbling on our lettuce, and also rabbits and groundhogs eating our brassica. We also have foxes, raccoons, skunks and a great population of toads, earthworms, and springtails. Who says we have no animals?"

8. The surplus animal population question. "Nationally, and globally, we have not only sufficient, but surplus farm animals. It is sad that they are mostly shut off from the natural world in factory farms, or in vast feed lots, but they do exist and they provide all the manure that could ever be needed for soil fertility. Why should we not bring this manure to use on our organic farms? We sanitize the land by composting these substances and use the compost thus to raise our crops and raise the fertility of our soils, instead of allowing them to leach into the ground."

9. The substitute substance question. "Out of various reasons I cannot keep animals but I use instead as much as possible the wonderful biodynamic preparations." This argument is oftentimes heard from biodynamic growers.

There are other arguments as well, but these are those most common. They are offered up alone, and in various combinations. If we cannot understand and discuss them properly, and show new ways of dealing with the problems they indicate, we will not have many farms in the future that integrate a variety of domestic animals into their organism. This will lead, as it is leading now, to a host of serious consequences.

There is an old and wise saying among European farmers: the pasture (grassland) is the mother of the field. This saying is based on many hundreds of years of farming experience, and it implies the understanding that if we do not have grasses (the mother) in the rotation, their fields (the children) will be lacking natural fertility. The grass roots are the main builders of organic substance—of humus—in the soil. This mother-child relationship is similar in human nutrition. Many physical dysfunctions and deformities in young people of today are based in the insufficient, denaturalized nutrition of the parents.

Humans would not have been able to enter earth evolution in their present form if ruminants, and before them grass, had not arisen in this evolution. Grasses and ruminants are preconditions of human evolution. The swamps and huge ferns of the early Tertiary period, with its huge dinosaurs, gave way more and more to mammoths, and then to grasses that made the lives of the ruminants possible.

In the late Tertiary period, through the cycles of evolution, there came first the goats, then the sheep, then the cattle, then the human beings. The earth had to lighten up out of permanent mist to allow the grass to come. Grass, ruminants, then human beings is one sequence of evolution, and the grass is the mother of the fertile field. This is hinted at in the Old Testament story of Noah. When the water of the great flood recedes, the first rainbow appears, for real light shines into the moisture.

The decline of humus and soil fertility all over the world, especially on the fertile prairie soils deriving out of the grass steppe, is a direct consequence of 160 years of the use of the modern plow and the use of incomplete one-sided rotations. When in the 1920s Russia claimed more and more of the grass steppes for cultivation and grain growing, soil scientists assumed that the layer of fertile topsoil was so deep that it would allow permanent grain growing. It took them only a few years to find out that the fertility was declining. They thought they could correct this by including grass clover lays in the rotation. They called this the *trava polnjanoa*—the grass-field system. But it did not work. The decline in fertility slowed but was not halted.

What was missing? The impulse of ruminants and their manure on the grass lays. We have to realize that the fertile, prairie soils are built up by thousands of years of grasses, with the help of many millions of ruminants—for example, the vast herds of bison which once roamed the American plains.

But it is not only the animal's manure that is needed. Farmers of old often said: "Sheep have golden hooves." Where sheep (or other ruminants) compact the surface of the soil with their hooves, without compacting lower layers, clover can be found surging. The Soviet collectivization of the farms did not allow the integration of the ruminant herds into integrated farm organisms. As in America, although out of different reasons, Soviet collectivization led to isolated animal production units without proper handling of the manure and without proper return of the manure to the fields. The consequences of this are obvious in both Russia and America. The degradation of the soils through one-sided plant rotation leads to the shrinking of organic matter in the soil, of topsoil, and of humus.

For vegetable growing on the poor soils of New Hampshire, we need to have two years of grass clover lays following two years of vegetables to rebuild the organic matter in the soil, and we need animal manure to fire up the grass clover lays—to feed the mother. The grass clover lay produces through its root system twice as much organic matter in the second year as in the first. At the same time, the clovers and alfalfa fix the necessary nitrogen from the air with the help of their root nodules.

The manure of the ruminants complements and enhances light forces in the gramineous plants, the grasses and grains. It is the key to crop quality.

Today a fashionable word in agriculture is sustainability. It is used to indicate the permanent upkeep of soil fertility independent of the source of this fertility. So we have organic farms today that are in a so-called sustainable system because they can sustain their fertility by importing feeds and organic manures from outside the farm. What does that mean? The fertility does not come back to the fields where it is taken from.

Separating crop growing and animal husbandry leads to the exploitation of the best soils of the world, to a general decline of fertility while the importing farm is in a so-called sustainable system. It is sustainability at the cost of other soils. Thus, the whole system is not sustainable.

We have a worldwide habit today that depletes soils to grow feeds for the mass production of animal products. The far too-large herds of domestic animals in Holland, Germany, Switzerland, and England, consume foods grown thousands of miles away. In a sense, they graze partly along the Mississippi, Missouri, and Ohio rivers of America. But they never give anything back to these soils where their grain has come from. Their manure is not returned to these distant fields where their feeds are grown, but instead goes raw, or often as compost, into organic farms, and that is said to keep these farms sustainable. True sustainability is only reached when all farms are sustainable through a soil-based fertility.

Here is an example from our neighborhood in New Hampshire: a certified organic grower grows two acres of vegetables in a beautiful setting. His livestock is three sheep—for decoration, one could say. He shows good vegetable crops and small fruit (pick your own). The fertility of his soil is based on sixty cubic yards of compost created from chicken manure purchased from a huge chicken operation. His vegetable operation is seasonal. In winter he works for the town where he lives, doing snow plowing, and driving a school bus. His wife works as a teacher. The questions can be raised: how are the chickens that produce this manure held and cared for? How are they fed? And how are the fields where the chicken feed is grown kept fertile? Despite the obvious answers, this farm operation would still be called, in the common use of the word, sustainable.

This leads us to the concept of the farm organism. A true and healthy farm organism produces all the manure it needs to uphold and raise fertility by keeping the necessary livestock. It has a true, indigenous production that does not come from imported feeds and fertilizers. Only here do we have full control over the quality of the manure and thus, inevitably, the quality of the crops that are grown in the soil fertilized by that manure.

Any organic matter can be composted somehow, but the forces in the compost depend on the forces that work in the excrements. When we keep our animals so that we feed them solely what grows on our farms, and ensure that it is adequate to the species, we start a process of individualizing the farm. The digestion of the farm's ruminants changes and adapts wholly to these local feeds and conditions. A ferment pattern builds up in the rumen that is in harmony with these feeds. The products of the digestive process are now in harmony with the local flora. They strengthen the local flora, and health and productivity is consequently built up. Horse manure mixed with sawdust, and filled with anti-parasitic toxins, cannot produce the right humus for crops. The dung from cows that are fed with twenty pounds of grain per day and more than 20 percent corn silage in the daily ration, is not only of a different consistency and an attractive breeding ground for flies, it is incapable in the fermenting process of developing into a quality of compost that helps to build the humus necessary for high-quality vegetables. That is because it is not just the physical, material nutrients that count, it is also the forces that are carried by these nutrients.

Because they rely exclusively on the local flora and in turn inspire the local flora with the forces of their manure, only domestic animals kept on the land are able to tie our farms into an organism. Yet this is not all natural. Rather, it is the result of voluntary acts by farmers to bring the substances and forces of the farm to a higher level of health and productivity. The imperative is not to go back to nature, but rather to raise nature to a higher level. The tool for accomplishing this is the man-made organism of the true farm, that individualizes the true farm. Individuality in the human realm is the cause of all progress. Brought down in the realm of nature, individuality causes health, stability, longevity, and productivity at the farm level.

Individuality is achieved when we are distinctly different from other beings, when our farm is distinctly different from other farms by setting borders that allow the substances and forces of the farm to develop to the highest possible degree with the least merging into the circulation of substances and forces from other farms. On the farm level, and on the human level, we need cooperation of individualities, not conformity.

Conformity in agriculture means worldwide plant and animal breeds that are poorly adapted to local conditions and therefore prone to diseases. For humans and for farms, the call of our time should be: individuality and cooperation, not merger, conformity, and uniformity.

A paper entitled "The Company in the 21st Century," published by a French association of 2,300 young corporate executives (Centre des Jeunes Dirigeants d'Enterprise), states: "For 20 years corporations have become profitable at the cost of society." We can modify this observation by saying: for fifty or more years corporations have been profitable at the cost of agriculture, the prime economic basis of society.

To overcome the raw capitalism underlying this process of exploitation we need a leading idea, or concept, that can bring us economically and ecologically into the future. Such an idea was first proposed by Rudolf Steiner in 1924: a closed, self-supporting farm organism, with a high level of life-giving forces, and a true (not borrowed) fertility and productivity. Based upon this stable and vital foundation of farms, the whole sphere of economics is changed.

Thomas Berry, Kirkpatrick Sale, and others have in recent years introduced the concept of bioregionalism. This concept rightly states that we need healthy bioregions to help solve our environmental problems. Miriam Therese McGillis, a Dominican nun and a CSA farmer in New Jersey, describes a bioregion as follows: "An identifiable geographic area of interacting life systems that is relatively self-sustaining in the ever-renewing process of nature. The full diversity of life-functions is carried out, not as individuals of species, or even as organic beings, but as a community that includes the physical as well as the organic components of the region. Such a bioregional community is self-propagating, self-nourishing, self-educating, self-governing, self-healing, and self-fulfilling."

From our perspective, a healthy bioregion can only be constituted out of healthy farm organisms, where every farm is, in a sense, a healthy bioregion. We can say: never will there be bioregional communities if we do not start at the farm level with its supporting community.

Applying to the farm the six life functions of the bioregion as Miriam McGillis describes them, we can say the following. The ideal farm organism should largely be: self-propagating, meaning that it raises its young stock of animals out of its own herd and raises seed for its plantings out of its own plants; self-nourishing, meaning that all its foods and feeds should be raised on the farm; self-educating, meaning that we take the fruits of learning from the close observation of the processes of the farm; self-healing insofar as the healing processes and agents for humans, animals, plants and soils should largely come out of the farm organism; self-governing insofar as all the social process among the people living on the farm and of the farm should be ruled by agreement of the people concerned; such a nature and social organism will then be self-fulfilling. The demands made here are ideals that never can be fully accomplished, but our striving should go in this direction. Such a bioregional farm, as the foundation of a healthy earth, necessarily includes and is carried in its fertility by domestic animals.

After an analysis of endangered animal species around the globe, the World Conservation Union stated that one-fourth of the animal species on earth are threatened by extinction, and that half of these extinctions may occur as early as the year 2006. The United States ranks among the twenty countries with the highest number of endangered species. The creation of self-supporting bioregions through self-supporting farms that do not use poisonous chemicals for their production, would be a great step toward the conservation of animal species, since the integration of domestic animals, and their health and the health of the plants, demands the planting of shade trees, hedgerows, windbreaks, the establishment of ponds—in short, landscaping that creates the right environment for all animal species.

Farmers have the foremost responsibility for our domestic animals which, for more than 10,000 years, have been our trusting and dependent companions. They have supported the human race and also made its evolution possible.

Now we have separated the animals from the human beings by putting them into huge compounds, and at the same time deprived vast lands of their beneficial influence. The degenerative consequences of this profit-oriented practice on the soils, the plants, and on the animals are only too obvious: deterioration of soil fertility through loss of humus, fungus diseases on the cultivated plants, parasitism, mastitis, and sterility in the domestic animals.

The need to reintegrate domestic animals into our farms, and to create species-adequate life conditions, is so urgent that nobody can bypass it. Still, all the reasons set out at the beginning of this essay for not keeping animals are good and valid. In the third part of this essay we want to address these reasons.

First, the land question: to support forty families with vegetables, including storables, one needs a minimum of two acres of land. To keep these two acres in a sound rotation, and to keep two cows, some sheep and some chickens, and also a pig to help clean up the garbage, you may need eight more acres. If this amount of land (ten acres) is not available, one might not be able to develop an integrated farm organism. If you are too close to a city the development pressure is likely to be so strong in any case that you might consider relocating further outside the city.

With eight to ten acres of land, one can establish a diversified, integrated farm organism with a variety of animals. In Europe, between the two great wars, when there was the first wave of a back-to-the-land movement, farm types were researched and developed that were called Gardeners Farms. Such farms, with well-kept livestock and year-round milk supply, can be very attractive to customers and friends who can much more easily identify with a small unit, than with a large farm.

Another possible option for a vegetable operation without animals is the close cooperation with a nearby organic dairy farm, and the structuring of both farms as one organic unity under the condition that you have at least enough land to rotate your vegetables properly.

That leads to the second reason commonly given for not keeping animals: the workforce and income question. An integrated farm organism is indeed too great a workload for one family, even if the farm is small. Two or three families should eventually live and work on a farm. They should, and need to have, two or three off-farm jobs to supplement the farm income. If we do not learn to live and work together as free, independent people, we will not have integrated farm organisms in the future, not even family farms, only agribusiness complexes—or factory farms— vast complexes for plant production, or for separate animal production.

Nobody should complain that he or she wants to farm, but has no land and capital. There are plenty of opportunities in the world for independent farmers. But if he or she comes to me alone or as a couple, I would say I cannot help you because a healthy integrated farm organism needs more than two people, if they do not choose to live a slave-like primitive existence.

This point also responds to the third question. Every farmer should be able to participate in the cultural life of his or her region, to have time for family and children, and to further his or her education at conferences and seminars. This is only possible if there is a team around the farm. The author has, as of this writing, been involved in farming for nearly forty years. He was able to attend most cultural and educational events he wanted to attend. But thiry of these forty years he had hardly any vacation. He did not even miss it, so much was he taken with his work. Loving farming, he did not feel the absence of vacations was a sacrifice.

The question of how free individuals (and families) can cooperate in favor of true diversified farm organisms becomes more and more urgent. Often it is not the insight into the necessity of such farms that hinders us, it is the lack of the social insights and skills necessary to keep up such an organism in cooperation with other individuals. We will eventually have to develop a new, modern village concept. Just as there is wisdom in the often-quoted saying, "It takes a village to raise a child," so it is true that it takes a village to develop and run good farms.

The fourth reason often mentioned for not keeping animals is vegetarianism, and the rejection of killing animals. This is an honest position. One could state that only 2 percent of the American population are vegetarians. However, this would be a weak argument because numbers are irrelevant in the face of such ethical questions. Yet there are farmers who eat a vegetarian diet, and still need and keep animals. Animal manure, properly handled and returned to the soil, gives us the highest quality of vegetable matter and the highest quality of vegetable matter is needed if you decide to eat a vegetarian diet. So we would have to advocate the keeping of domestic animals, which includes eliminating old, weak, and surplus animals, even if no one would eat the meat.

I believe that in the future more people will turn away from meat eating, but they will still need and want milk. Certified organic/biodynamic raw milk is essential and not replaceable for the upbringing of children, and in processed form for the adults (see *The Milk Book,* by William Campbell Douglas, MD).

For many people the killing of animals is a delicate, divisive problem. In some cultures—Jewish and Muslim, for example—butchering is a religious task, and with good reason. But, in general we have to fight for humane and local facilities for butchering. To understand what is involved here we would have to understand what the meaning of death is in the animal kingdom, compared with death in the human realm. In no case whatsoever, should we inflict unnecessary pain upon animals.

Part of this problem is the question of whether we can replace animal manure totally with green manure—plant matter compost and mulch. I have serious doubts that this is possible. Permaculture is an interesting option, but is unlikely to ever meet the needs of humanity for grains and other crops.

The sixth reason cited against keeping farm animals is hygiene. Some people argue that manure has a bad smell and attracts flies. However, experience teaches that it is not the animals, but their masters or mistresses, who are the cause of bad smells and excessive flies. If animals are fed properly according to their particular nature, neither they nor their excrement smell bad. This is true even for pigs and pig manure. And if we treat animal manure properly after it has fallen, and bring it into a mild fungus fermentation, it does not attract flies. What does not smell offensive is unattractive for flies.[1]

The seventh common reason is the belief that the animal element is sufficiently present on the farm through all the wildlife—from deer down to the earthworms. We have pointed out that the dynamics of the farm are enhanced by the strong inner circulation of feed that goes to the animals that then give manure, that is then recycled to the fields, and thereby stirs new plant growth. This dynamic we can achieve and control only with farm animals. The non-domestic animals roaming the farm are welcome and important additional helpers.

The eighth reason for not keeping animals on the farm points out rightly that overall there are too many farm animals in the world, and that we should use the surplus manure produced by this large population of animals on the organic farms we are tending. Here one has to state that the quality of "animal factory" manure is dubious, and that there is good reason not to support this inhumane system of animal keeping by purchasing the manure produced on such factory farms— thereby helping such operations to eliminate what for them is a major problem.

Finally, the ninth argument, that one could replace the use of animals and their manure by an intensive use of biodynamic preparations and sprays, is, to me, nonsensical. We cannot replace sound farm practices—that include animal husbandry—with any preparation. The biodynamic preparations and sprays are only an additional help to basic, sound, time-tested farming practices.

The reasons for farming without animals are mainly socio-economic, rather than agricultural. As pointed out by Sir Albert Howard, and before him by Rudolf Steiner, a real farm is not possible without animals. The socio-economic reasons such as lack of land, lack of help, lack of sufficient income, cannot be dealt with by the small number of farmers alone. They need the help of many non-farmers who are concerned about ecological and nutritional questions. This is the basis of community support for farming. To guide the interest of non farmers to the farms of the future requires a leading concept. This leading concept is the self-supporting, diversified farm organism with a great variety of plant and animal species. This is not just an economic, but also a cultural-spiritual idea.

People are attracted and held together not by consumer egotism, but by ideals. My ten years of experience in a diversified community farm that supports 105 families year round with produce, meat, milk, and eggs, has given me that conviction.

1. Nicholas Remer. *Laws of Life in Agriculture.* Kimberton, PA: Biodynamic Farming and Gardening Association, 1995.

Essays on the Context and Scope of
The Farms of Tomorrow

Steven McFadden

The Context of Community Farms:
Industrial Culture and Agriculture

The many encouraging initiatives of Community Supported Agriculture (CSA) arise not in pastoral isolation, but rather amidst the vast and jangling context of global industrialization. The modern industrial processes of efficiency and mass scale have been brought to bear not just upon factories, but also upon a wide range of human activity, including our farms and our food.

Since the first edition of this book in 1990, agricultural industrialization has proceeded at an undiminished gallop. Thus, for the understanding of people interested in or involved with community farms, it is important to illuminate this global context, and the host of related issues. Global industrial culture constitutes the larger environment in which community farms will either thrive or perish.

While it may have the ring of an ideologically skewed neologism, the term "industrial agriculture" is not new. It has been around at least since the late 1800s, and refers generally to the process whereby agricultural production becomes less a way of life, and more a commercial activity. While industrialization has been most often associated with crops grown via chemical techniques and livestock raised with quantities of antibiotics and hormones, it has already begun to influence, and may eventually come to dominate, what is known as organic or natural farming as well.

A report entitled *The Industrial Reorganization of US Agriculture,* from the Henry A. Wallace Institute for Alternative Agriculture (April, 1996), sets out the generally accepted characteristics of industrial agriculture: an expanded size of operation, an increase in the use of chemical fertilizers and pesticides, an ongoing drive to replace labor with technology, increased use of cost accounting, a stress on uniformity, a narrow genetic base for the majority of major crops and livestock, and a decrease in the number of owner-managers.

Meanwhile, a 1996 report from the USDA's Advisory Committee on Agricultural Concentration reported on the tendency of industrial agriculture to concentrate agricultural resources in fewer and fewer corporate hands. "We find these issues complex and highly charged," the committee wrote, noting its concern about the balance of economic power, the use of government power, and issues of personal freedom. The USDA report explored in detail several consequences of this concentration: distorted prices, unequal access to market information, dysfunctional interactions between producers, handlers, and others in the food-to-consumer chain, and environmental degradation.

What is happening in agriculture is but a part of general trends in economic sectors around the globe. As reported by the United Nation's Development Program (UNDP) *1996 Human Development Report,* the gap between the rich and poor is widening every day. Globalization contributes to this, the report finds, and has the possibility of becoming "a monster of gargantuan excesses and grotesque inequalities." Considering a host of facts, statistics, and human lives, many observers believe that this has already happened.

In some respects, what follows may appear a tedious onslaught of facts. Together, though, these facts render a picture that every citizen should be aware of. Global industrialization is as much a part of the agricultural scene today as it is part of fishing, logging, banking, manufacturing, and health care. This is the way it is. The consequences of the agricultural-industrial complex—not just for farmers, but also for families—are crucial, and they are just slowly coming into pubic focus. Like it or not, the consequences are part of our reality every time we sit down to eat.

Disturbing Trends

As of the mid-1990s, we are still in the farm crisis. It didn't end when the 1980s ended, it just evaporated from the front pages. While population continues to grow, family farms continue to disappear. Consider these statistics from the US Census Bureau:

- 1900: 22.9 million people living on farms, almost 40 percent of the population.
- 1980: 6.5 million people living on farms, 2.7 percent of the population.
- 1990: 4.75 million people living on farms, 1.9 percent of the population.
- 1993: The Census Bureau drops its count of the number of citizens living on farms. The numbers are so small it is no longer worthwhile to collect data.

In 1994 the USDA reported that the nation was still losing an average of 23,000 farms each year (about 1 percent). Most of these are small farms which no longer produce sufficient income to support the family. According to the USDA, 85 percent of the nation's agricultural output now comes from about 15 percent of the nation's farms.

While these statistics do not make the point directly, it is also true that instead of owning the land and having an intimate, generations-old relationship with it, farmers are increasingly becoming hourly workers for large corporations. They are specialists, their work is hard, and they have no investment in the operation. They don't own the land or even have a stake in it. Under these conditions, hourly workers have scant motivation to caretake the land for the generations to come.

Meanwhile, compelling evidence from around the world suggests that current farming practices in many areas cannot be sustained much longer. Conventional economic indicators used in the agricultural sector neglect measures of environmental damage, and such evidence rarely makes its way into economic decision making. Indeed, agricultural sustainability—though broadly recognized as important—is given little weight in economic policy-making. No commonly employed indicators measure it, no accepted conventions value it, and no widely accepted definition describes it. If agricultural sustainability is considered at all, it is an afterthought.

For family farms, the prognosis is not good. Obviously America is still producing a lot of food. Where's it coming from? Increasingly from large corporate enterprises that, through expansion and merger, continue to emphasize technology, standardization, efficiency, and profit. As has been true for decades with wealth in general, in agriculture resources are being steadily concentrated in the hands of a few. The antithesis of democracy, the overall trend of industrial agriculture has disturbing consequences:

- In 1978, 50 percent of all market value for agricultural products sold was accounted for by the largest 130,000 US farms. By 1992, according to the California Institute for Rural Studies, just 61,000 farms accounted for 50 percent of all market value. This trend has continued in subsequent years.
- Industrial farming practices result in the devastating loss of topsoil, and they are the source of a staggering degree of pollution of the air, the water, and ultimately the food. According to the World Resources Institute, more than three billion acres of land (an area the size of China and India combined) have been seriously degraded

since World War II. The cause of this degradation is primarily chemical fertilization, high-tech cultivation techniques, and livestock overgrazing.

- Every year corporate farms apply to croplands an estimated 850 million pounds of pesticides and herbicides, plus over 22 billion pounds of petrochemical-based fertilizers. Much of this toxic material eventually finds its way into the waters.

- Nearly 70,000 chemicals are used in various commerical activities, including many pesticides employed by various industrial agricultural concerns. For many years US pesticide laws were based on "risk assessment," and there were still an average 10,000 pesticide-related cancer deaths each year according to *Food & Water Journal* (Summer, 1996). New US pesticide laws passed in 1996 are likely to be even less effective in providing protection to consumers and workers. Meanwhile, a study published in *Science* magazine in the spring of 1996 shows that while single, isolated chemcials may not be carcinogenic at the levels usually studied, combinations of two or three common pesticides, at low levels, are up to 1,600 times as powerful as the individual pesticides by themselves.

- Between 1982 and 1990 there were 4,100 food industry mergers and leveraged buyouts (debt purchases of another firm's stock). At present there are only a handful of major package goods companies in America, including the so-called "tobacco giants" RJR Nabisco and Philip Morris. The major brands of products that fill our supermarket shelves—while they may all seem to be different—are really almost all from these companies.

- Many agricultural corporations are going private to escape government and public accountability.

- Technology and chemical-based industrial farming is capital intensive, and depletes resources. It contributes to social inequity because it caters to those who have the resources or power to access capital and credit. As a result, it worsens poverty in the countryside.

- Approximately eight billion animals are raised for consumption each year on factory farms in the US. In many instances, these animals are raised in dark, squalid confinement. The dumping of millions of tons of animal waste and rotting body parts from these massive operations frequently poisons waterways. And under the harsh conditions of industrial efficiency, millions of the creatures raised for human consumption perish long before their scheduled journey to the slaughterhouse.

- Food security is at risk because the amount of fresh water that can sustainably be supplied to farmers is nearing its limits, even as 87 million more people are added to the planet each year. Industrial agriculture is a major consumer of finite fresh water supplies. Within a generation, this issue of food security will be brought into sharp relief, according to The Worldwatch Institute's 1996 report, *Dividing the Waters: Food Security, Ecosystem Health, and the New Politics of Scarcity.*
- While industrial agriculture grows plenty of food at present, the bottom-line mentality it is tied to results in millions of people going hungry. The commercial-industrial system is geared to getting food to those with the most money to pay, not those with the greatest need. According to the *Hunger 1995* report from the Bread for the World Institute, at least 800 million people, mostly women and children, are chronically undernourished. Every day 40,000 people die of hunger and hunger-related causes. The problem is not just overseas. In the US, thirty million people suffer chronic undernourishment. Almost half of the hungry are children.
- People who labor on farms still earn the least amount of money working the hardest jobs with the worst living conditions of any other group of workers (*The New York Times,* 7/9/96). They often work twelve to thirteen hour days, and at the large, industrial farms they are regularly exposed to dangerously toxic chemicals. Farm workers under this system are often treated as migrant commodities, rather than as human beings. In general, the living conditions of the people who grow our food are appallingly substandard.
- More than 100,000 children are working illegally on US farms, according to the National Child Labor Committee. And according to the Congressional GAO, between 1979 and 1983 approximately 23,800 children and adolescents were injured on farms, and 300 died. Up to 40 percent of these children have been sprayed with pesticides.
- As a solution to the nation's farm problems, the USDA has been recommending an emphasis on biotechnology (genetic manipulation of seeds, plants, animals, and microorganisms) even with all its long-term uncertainties.

Where is the Juggernaut Headed?

Where are we going? What are the farms of tomorrow going to look like? Who's winning? Who's losing? What is likely to happen? The following observations begin to form a picture:

- World demand for food is being driven relentlessly upwards by the annual addition of 87 million people. Merely maintaining the current world per capita grain consumption requires and additional 28 million tons of grain production per year, or 78,000 tons more each day.

- Massive food shortages may well develop over the next forty years as the ongoing population explosion outstrips the world's food supply. In the Worldwatch Institute's 1994 book, *Full House: Reassessing the Earth's Population Carrying Capacity,* the authors write that "the food sector is the first where human demands are colliding with some of the earth's limits: the capacity of oceanic fisheries to supply fish, the availability of fertile new land to plow, and the ability of the hydrological cycle to supply irrigation water." They conclude that "food security will replace military security as the principal preoccupation of national governments in the years ahead."

- Agricultural Goliaths are definitely on the rise. As Chuck Hassebrook of the Center for Rural Affairs has argued, federal farm policy has been biased in favor of bigness, and has subsidized the use of capital to replace people. Federal policy has consistently fostered the industrialization of agriculture, including the concentration of agricultural assets into fewer and fewer hands. His points were echoed in a series of *New York Times* articles (Oct. 10-12, 1993), documenting the incestuous relationship between the USDA and agribusiness interests.

- A 1993 study by the World Resources Institute shows that most of the world's industrial agricultural practices are unsustainable. The way things are now, and the way they are developing, we're going to continue to loose soil, water, and farmers, all while population is skyrocketing.

- Hormone-impregnated, irradiated, and genetically altered foods will increasingly appear in grocery stores. Dr. Lowell Catlett, an agriculture specialist at New Mexico State University, has observed that at least 110 new genetically engineered plants and animals will be released over 1996. Before the end of the decade, he says, we will see 500 new genetically engineered plants and animals. Corporations have been investing over $40 billion in bioengineering research over the last ten to fifteen years (much of it funded with public money), and more than $7.5 billion in just 1996. We are just entering what these investors and companies hope will be their "payback time," when the commodities will start showing up on supermarket shelves—in all likelihood without labels identifying them as genetically altered.

As industrial fishing and factory trawlers have driven the boats of fishing families from the sea, so industrial agriculture and factory farms have driven—and continue to drive—farm families from the land. It's a different medium, but the same process.

Consider these names: Agricetus, Nanotronics, Metabolix, Monsanto, Millennium Pharmaceuticals, Genomyx, Calgene, Dow Chemical, Genetech, Unilever, Archer-Daniels Midland, and RJR Nabisco. If we continue the way we're headed, these are the farms and farmers of the future.

The Global Specter of Codex Alimentarius

All of these diverse yet interrelated trends are overshadowed by the global umbrella of Codex Alimentarius, an organization most farmers have never even heard mentioned.

Back in 1962 the Food and Agriculture Organization of the United Nations and the World Health Organization called for food standards and guidelines that would "protect consumer's health and ensure international fair trade practices." As a result, an organization called the Codex Alimentarius Commission (CAC) was formed.

The term "Codex Alimentarius," taken from the Latin language, means "food code." In essence, Codex is a set of international standards for food, including codes of manufacturing practice, that the organization says protect the health of consumers and reduce unfair practices in international trade.

With the adoption in the early 1990s of the Global Accord on Tariffs and Trade (GATT), the Rome-based Codex Alimentarius Commission became an even more important—if not more visible—international bureaucracy. As of 1996, Codex Alimentarius has 151 member nations. The Commission has produced twenty-eight volumes of standards, guidelines and principles, including 237 food standards and forty-one hygienic and technological practice codes.

The public presentation of Codex suggests that it is an international device that will make all countries play by the same rules, encourage the most efficient producers, and reduce the cost of farm programs. But many observers believe Codex Alimentarius amplifies and accelerates agricultural policies that have already proven destructive of farms, farm communities, farm culture, and food.

By any estimation, Codex is remote, and for the average consumer unknowable and difficult if not impossible to influence. In 1995 attorney Suzanne Harris, representing the Life Extension Foundation, reviewed a list of international organizations allowed to send delegates to Codex. She found that more than 90 percent of them represent multinational pharmaceutical corporations. The general public has virtually no representation.

Codex Alimentarius gives decisions made on a remote international level potent influence over member nation's internal affairs. In a sense, they mandate the sacrifice of local and national sovereignty over basic choices about food. Any nation which does not accept the Codex standards can be heavily sanctioned and fined by the World Trade Organization (WTO).

Codex can be seen as tending to exacerbate the trends developing in agriculture over the last thirty years—emphasis on increased production, high efficiency, lower prices, and technological advantages—smoothing the ongoing process of global food industrialization. While farmers and the public are generally uninformed about how GATT, WTO, and Codex impact them, they do have an impact—an impact that will be increasingly felt. Many troubling issues and proposals have come before these interrelated, international groups, and such issues will continue to arise:

- One GATT provision allows DDT levels to be fifty times higher than the US currently allows on imported peaches and bananas—and twenty or thirty times higher on most everything else.
- Under the *Food Quality Protection Act,* the US Environmental Protection Agency (EPA) would be required to justify any pesticide residue tolerance that is different than those set by Codex Alimentarius. Tolerances set by Codex are often much more lenient than EPA standards.
- The Codex Alimentarius Commission adopted recommendations in 1995 for synthetic "growth-promoting" hormones used in food production, despite a European Union ban on the use of these substances. Veterinary drugs which promote growth in animals are used in major meat-producing countries such as the United States and Australia. Codex considered it unnecessary to establish Acceptable Daily Intakes or Maximum Residue Limits for these hormones.
- The German delegation to Codex has proposed radical changes in the rules governing dietary supplements which, if passed and implemented, would require a doctor's prescription for the vast majority of natural dietary supplements now available. The plan calls for the following: no dietary supplement could be sold for preventive or therapeutic use; no dietary supplement sold as a food could exceed

the potency levels set by Codex; Codex regulations for dietary supplements would become binding; all new dietary supplements would automatically be banned unless they go through the Codex approval process.

- The Codex Committee on Food Labeling met in May, 1996 in Ottawa, Canada to debate whether genetically engineered foods should be labeled as such. If the biotechnology initiative is enacted, consumers will be unable to identify genetically engineered foods in the market and, thus, will be unable to make an intelligent free-will choice concerning possible risks associated with these largely untried additions to the food supply. Consumers will never know what they are purchasing.

These are just some of the issues that have come under the umbrella of GATT, WTO, and Codex Alimentarius. Over time, especially now that the organizations and their regulations are more fully empowered, vital issues will continue to arise. In most cases, it is unlikely that small-scale farmers, or the general public, will ever be aware of the issues, the process, or the decisions. But all will ultimately experience the consequences.

This is a crucial concern because, in a very real sense, trade and agricultural policy are also social policy. They determine not only what most people eat, but also where and how people will live, how the land is held and worked, and the character of enterprises that contribute to food production and processing. Trade and agriculture policy can contribute to a regenerative strategy for rural community development, or they can also undermine it. GATT, WTO, and Codex are mechanisms that emphasize large-scale and "efficiency" to a point where multi-national corporations are favored, usually at the expense of smaller, less profitable businesses, including family farms.

John Hammell, coordinator of the Life Extension Foundation and an active observer of Codex, has written that he sees the international agency as nothing less than a vehicle of "commercial colonialism," through which the world's largest food, chemical, and pharmaceutical corporations are laying claim to what has been the traditional territory of farmers and the health food and supplement industry.

Their increasing hegemony, he suggests, is reminiscent of the colonists who once laid claim to new territories in Africa, the Americas, and elsewhere. By simply claiming new territories as their own legally, the colonists began eliminating the the indigenous populations and eventually pushed those who survived onto reservations that the colonialists also defined. This time the territory is food, herbs, and vitamins.

In his 1994 book *Sex, Economy, Freedom and Community: Eight Essays,* Wendell Berry argued passionately that there is no doubt about the undermining effect of these global mechanisms. He wrote that the GATT agreement subordinates the interests of all governments, along with the people they represent, to the will of a few supranational corporations running a so-called global free market. "We must see that it is foolish, sinful and suicidal to destroy the health of nature for the sake of an economy that is really not an economy at all but merely a financial system—one that is unnatural, undemocratic, sacrilegious, and ephemeral."

Crucial Questions, Dubious Diet

For the foreseeable future, these global agricultural trends are likely to continue. Seen in this context, independent, human-scale farms are a fragile treasure—an endangered species in America, and around the globe. Since farmers are the stewards of our civilization's foundation—in a sense our ambassadors to the Earth—the consequences of this shift are profound indeed.

When the facts and trends are set out plainly, they begin to bring some fundamental questions into focus. What is the purpose of food and farming? And in this context, what is the purpose of knowledge, information, and technology? Those questions invariably lead to others: How strong is your faith in the banking system? The real estate system? The chemical corporations? The biotechnology industry? The government? By any estimation these are now the prime factors in the industrial-agricultural equation. Though seemingly remote, they do matter.

At industrial farms and as it is handled by large corporations, food becomes a commodity, reduced to the physical properties of minerals, vitamins, and calories that can be precisely synthesized and manipulated. Through a focus on profitability, the soul of the land is excised; and through petrochemicals, synthetic hormones, genetic manipulation, irradiation, and a host of other dubious materials and processes, the basic character of our food is altered. Fruits, vegetables, grains and meat become nutritionally equivalent "food products." The unacknowledged philosophy of materialism holds sway from seed through sale.

John Robbins presents a stunning array of facts concerning the particulars of this materialism in his popular book, *Diet for a New America.* He writes, for example, "today's factory farm livestock are subject to vast quantities of toxic chemicals and artificial hormones. . . . Over 65 percent of the antibiotics produced in the United States are not used as medicines; they are used as feed additives on factory farms. . . . Hardly any of these chemicals even existed before World War II, and so we have yet to witness the longer-term health consequences of eating the products of the factory farms, which invariably contain residues from pesticides, hormones, growth stimulants, insecticides, herbicides, antibiotics, appetite stimulants and larvicides."

A Danger For CSA: Marketing Is Not Community

The demise of the family farm has been a long-playing melodrama. It is in its final scenes, though not many citizens have noticed. Independent farmers endure as advertising icons, such as the romantic image of the Jolly Green Giant, or Juan Valdez and his burro harvesting Colombian coffee beans. But these are cynical illusions. The reality is increasing industrialization and globalization. Through these processes, steadily, more and more power is transferred to supranational corporations, and managers, rather than elected officials, increasingly become the policy makers. So remote is this process that citizens are virtually unable to learn about, never mind enter, the debate about the consequences.

For sober observers there is not so much the sense that this is some sort of intentional conspiracy—as if men were sitting around darkened rooms consciously plotting how to incrementally poison and dispossess the masses. Rather, there is a recognition that, no matter what the conscious aims of global deal makers, these are the consequences—consequences that, if they see them at all, are apparently inconsequential to them. While the corporate-industrial form is not malign in and of itself, as practiced at large in the world it is surely not imbued with a wholesome community spirit.

As industrial agriculture continues to alter the basic character of civilization's foundation, we are in need of a collective pause to examine the motivation and the consequences of our farming practices. We are also in need of thousands upon thousands of people who can grasp and act upon what might be termed "more reverent alternatives." While farmers must be among these leaders, they cannot stand alone, outside the communities where they toil. They must have the active involvement and support of their neighbors in many different ways. CSAs offer one range of possibilities for this to happen.

Jean Yeager, a long-time observer of CSA, and former managing editor of *Biodynamics,* the journal of the Biodynamic Farming and Gardening Association, regards CSA as a parallel polis. He explains this by citing Vaclav Havel, former president of Czechoslovakia. In a series of essays entitled *Living in Truth,* Havel pointed out that while there may be a powerful status quo in any given sector of society, there can also be a group of people living alongside or within that status quo who adhere to other values and other pursuits—a parallel polis. That's what CSA is now, Yeager observes, a parallel food system to the status quo of industrial agriculture.

At this stage, in the late-1990s, global market forces are slowly beginning to perceive that there is profit to be made from the "natural market," and thus we are seeing more and more the development of an "industrial organic agriculture." In time, if more people become educated about toxins in their foods and choose organic foods in the marketplace, chemical-industrial agriculture could well be replaced by organic industrial agriculture. For some, it's just another market to exploit. And these forces may even come to bear upon the CSA initiative.

"Marketing specialists have identified a hunger in the human soul for contact with the land and for community. They have set up an artificial system to meet that hunger," Yeager observes. "In a way it's sad how they have done that. The hunger in people's souls is genuine. The hunger is not manufactured, but the response often is—and it cannot possibly meet the true need." He points out by way of example that people living in the West can now go to "pick" pumpkins locally. The pumpkins are not grown locally, but rather trucked out into local fields from where they are grown, many miles away. Teams of local students scatter the pumpkins over the field. When families come for "U-Pick Pumpkins" in October, they walk out in the field and "harvest" them. The experience has been synthetically replicated to meet the market for authentic contact with the farm and nature.

"A danger for CSA," Jean Yeager observes, "is the possibility of this widespread, soul-felt yearning for a connection with the earth and community being exploited to the extent that CSA is perverted, becoming more and more a marketing exercise, and less of an authentic community enterprise."

As ludicrous as it may sound, there could someday be a nationwide CSA franchise—Olde McDonald's CSA, for example, with actors dressed as 'Olde Farmer McDonald,' wearing neatly pressed overall and checked-shirt uniforms, handing out baskets of berries and bananas to the patrons and their children. It depends on whether large food corporations ever see the potential for profit; and on whether this is an option consumers would embrace. It could happen.

But marketing is not community, and merchandising is not CSA. CSA is about a direct relation to the Earth and the people who work the

Earth on your behalf. To the extent that this is understood and embraced, CSA will continue to thrive, whether as a parallel polis, or as a widely understood and appreciated part of the world economy.

"Your personal food choices directly affect the world and the environment," Jean Yeager insists. "If you buy some kind of fast food, you cause a certain use of the Earth to happen—factory farming, for example. If you don't buy the fast food, but instead support a CSA, you know precisely what you are doing not 'to' the Earth, but 'for' the Earth. We can easily forget that reality. With our choices of food, we each directly affect the Earth. Consumers are ultimately in control. If they support industrial agriculture, it will continue to proliferate, and so will all its consequences. And if they support CSA, it will continue to prosper and proliferate, and so will all of its consequences."

Essay 2

Belonging to a Community Farm:
Families, Food, Farms, Festival

Like reeds in a basket, human life is interwoven with the life of the earth. All our food, water, clothing, and shelter come from her body, and arise with her natural rhythms. Our skin and bones are likewise formed of her stuff. Our moods, thoughts, and capacities are not wholly independent of this relationship.

As earth has a rhythm, so do all the creations which inhabit her. Those rhythms are related. To lose or forget this relationship—as many people have in their modern lives and diets—is to invite discord into the individual life, and eventually into community life as well.

Human response to the basic rhythms of the planet is plain enough. We have definite, measurable reactions to fundamental cycles such as day and night, and the seasons, and these responses are well charted by medical technicians. Yet at the same time there are subtle inner rhythms in both planet and person, rhythms that mirror the level of either harmony or discord in a life.

A rhythm is a cyclic process that, by definition, is connected not to a theoretical or artificial division, but to something real: the waxing and waning moon, the earth's revolution around the sun, our in-breath and out-breath, and even our digestion—which begins with the rhythm of chewing and progresses through peristalsis.

When we fall out of harmony with nature's rhythms, there is a conflict, or arrhythmia, that eventually manifests itself in our lives in a perceptible way as a physical malady or a psychological disturbance.

One gift that CSA gives to individuals, to families, and to culture in general, is a vehicle for re-establishing a conscious connection with the rhythm of life, the rhythm of the seasons, and the rhythm of the farm that gives rise to the food which eventually becomes the molecules and cells of our bodies. Thus, joining a CSA is an act not just of economy or ecology, but also of health at its most fundamental level.

Burdens and Blessings

Although everyone eats, fewer and fewer people have any personal ties to agriculture. And yet because we eat, we are all ultimately involved in and dependent upon farming and agriculture. As poet, essayist, and farmer Wendell Berry has observed, eating is the last act in the annual drama of the earth that begins with farmers and planting. What ends up on our dinner plates is directly connected with the lives and the activities of farmers and their farms.

For most people, for most families, this fundamental dependency on farming seems remote. If farms and farmers are thought of at all, they are taken for granted. There has always been an abundance of food, it seems, and there is no need to expend time thinking about it, other than to draw up a shopping list for the next trip to the supermarket.

Yet many parents have had the experience of discovering that their children are under the impression that meat and vegetables magically arrive from factories, fresh-frozen and cloaked in brightly colored cardboard boxes, ready to pop into the microwave oven. It's a logical assumption in the absence of other insight. A few years back a major American dairy group surveyed school children in an effort to learn what they knew about food. They were astonished to find that many of the children had no notion of the relationship between a cow and glass of milk. In fact, many of the children believed that milk came from a factory. How were they to know otherwise? They never see cows. Instead they see milk in stores, or in their family refrigerators, lined up in handsome, sanitary packaging and displayed in much the same way as sodas and other manufactured beverages.

This anecdote raises fundamental questions for everyone: What do we know about our food supply? About its quality, its wholesomeness, the earth the food is grown in, the lives of the people who grow it, and the complex systems that produce it and eventually bring it to our kitchen tables? For most people in modern societies, these are unfathomed mysteries. The natural rhythms that give rise to our sustaining food and fiber are unknown, unacknowledged—and thus seemingly irrelevant.

In most cases CSA brings those rhythms and realities—and their fundamental importance—back into focus. Membership in a community farm provides a direct link to food production that is impossible for shoppers who rely on supermarkets or even farm-stands. CSA members see their vegetables planted, watch them grow and ripen, sometimes even get dirt under their fingernails helping in the gardens. They develop a personal relationship with the farmer and the farm. Their children learn first hand about the source of their families sustenance.

But CSA is not for everyone. It's not always fast, easy and efficient—the qualities most of us are accustomed to through the modern marketplace. CSA takes commitment. Participation is a huge thing to add to your personal or family life. Especially for new CSA members, there are a range of concerns:

- Members have to pay money up front—from $175 to $600, depending on the CSA and the size of the share—for vegetables which are not even planted yet. This is difficult for some families, both financially and emotionally.
- Members have to pick up the food at the farm or a central drop-off point, and while there interact with the farmers or other CSA members. Some CSAs do deliver for an extra charge, but the preferred method of disbursing food is a central drop off. Not everyone likes this.
- Some people are put off by CSA because they feel they get too much food. They are neither accustomed to this bounty of fresh, raw foods, nor educated about it. A common complaint heard by CSA farmers: "This is too much. I don't know what to do with it, and I don't want to waste it—because then I will feel guilty."
- Some of the vegetables that CSA members receive as part of their weekly share may be new and unfamiliar. Not everyone has experience with collards, beet greens, and so forth—and they may be a hard sell for finicky children. Getting people to broaden their diet from the four or five basic vegetables they are now accustomed to, thanks to mass production, is a real tricky thing. People know iceberg lettuce, tomatoes, peppers, and onions, and that's about it. Dietary habits are set deep and, naturally, of great emotional importance to each individual. Consequently, there can be lots of personal resistance to different ways of eating.
- Members receive local produce seasonally. Therefore, naturally, there will be no melons in March, and no lettuce in February. This accommodation to nature's rhythm and reality comes as a shock to people acculturated to the year-round bounty provided by industrial agriculture. In the supermarket, the concept of seasons seems almost a quaint notion.
- CSA membership tends to force families to cook and eat in more. They have to do something with the fresh food that has arrived as their weekly share. Thus, member households may need to learn new ideas about food, and new ways of cooking.

- Some shareholders complain that they must take whatever food is offered each week, even items they would not choose to buy. It can be perceived as a series of unnecessary headaches: picking up the food, cooking it, eating what is available in season from the farm, rather than the vast array of exotic choices imported from around the world and available at the local supermarket.

While each of these points may be viewed as a problem by some new CSA members, others will see them as strengths or advantages.

Louise Frazier, an author and an experienced chef, has observed the development of CSA for nearly a decade. She believes that the reason CSA growth is strong but not explosive, is that members have to do inner work—to make an inner shift as well as an outer shift. People can, and most often do, choose to go to an anonymous supermarket and buy fruits and vegetables that they know nothing about. Or they go to a farmstand and buy directly there, but have no relationship with the farmer or the farm. When people join a CSA, she says, they are entering into a direct but complex relationship with the earth and other people. They are purchasing food which is 'informated,' or laden with a particular kind of information and value. Not only does the CSA grow food for your family, but the operations of the farm are, theoretically, in concert with your values. "Your purchase of food makes it directly possible for someone to care for the earth," she observes. "It really is about a lot more than vegetables."

Above and beyond any other motivation, it is apparently the quality and variety of the produce that lures most people to join a CSA and keeps them interested year after year. CSAs offer foods that would never survive being picked at the peak of perfection, being trucked long distances, and then being pawed over by shoppers.

Other elements also have a strong bearing on member loyalty and participation. Since community farms typically use organic growing techniques, members' concern over chemical residues is alleviated. Members share recipes, discuss eating foods in season, and enjoy cooking and eating high-quality food. As a consequence, some CSA members have shifted to eating far healthier diets simply because the food is there, readily available to them each week. This reality—this deliberate move to be in harmony with the rhythm of the seasons and the farm—can naturally improve their health.

Many CSA members say their participation has led them to try new things that are infrequently found in supermarkets, foods such as kale, sorrel, daikon, and kohlrabi. While this variety has caused some families to drop their CSA membership, it is welcomed and praised by others. The rhythm and the variety typical of CSAs is thus both a blessing and a burden.

Variety is much more likely to be recognized as a blessing when CSA members are educated about the food they are receiving. In recognition of this reality, some CSAs distribute a one-page farm letter each week, with a short note from one of the farmers on one side, and recipes for that week's most abundant or unusual vegetables on the other side. CSA members are often as keen about getting their farm letter and recipes as they are about getting their vegetables. Yet other CSAs have developed books or booklets to educate members about the foods, and how to prepare them (see *Appendix F—Resources*).

One book, *Louise's Leaves: Around the Calendar with Local Garden Vegetable Produce,* written by Louise Frazier, strives to educate people about the wide variety of vegetables they will receive in a typical CSA share, and to offer tips on preparation. Likewise, in a series of articles published in *Biodynamics,* the journal of the Biodynamic Association, she makes important observations on nutrition, using the food a CSA provides, and eating in harmony with the seasons.

Harmony with the Seasons

Louise Frazier recommends that we strive to maintain a diet based primarily on in-season foods that are grown in the region where we are living. Following this natural principle supports the body, helping it to be in harmony with the rhythm of the season.

The natural rhythm of the seasons has been all but forgotten in the global marketplace. For millennia, human beings were accustomed to and adapted to eating local, in-season foods. There were no other options. With the advent of industrial agriculture and the global marketplace, our diet has shifted radically—we can have formerly exotic foods at virtually any season in virtually any location. But what are the consequences of this radical shift? They are rarely considered.

Local foods naturally help us to adapt to the local environment, and in-season fruits and vegetables naturally help us adapt to the local climate and its rhythm. Such foods invariably have superior vitality and flavor. Eating in harmony with the seasons, Louise Frazier believes, can be a point of departure for culinary creativity and discovery—an opportunity to learn and explore approaches to cooking and eating that are inherently life-enriching.

She observes that when people base their winter diet on root crops such as carrots, beets, turnips, and potatoes, the body has a natural tendency to be warm despite cold weather; whereas if the winter diet is based on out-of-season foods, such as melons, zucchini, eggplant, and peppers, the body may have a tendency to feel cold.

In the dead of winter many people eat tropical fruit from southern climes, tomatoes from European hothouses, and strawberries from Mexico. Oranges, onions, figs and olives arrive at our tables from thousands of miles away, at all times of year. It is a bounty of legendary proportions. But this bounty consumes vast amounts of petrochemical energy for fertilization and transportation—the average meal in America has traveled well over 1,000 miles before it gets onto a plate in front of the person who will consume it. Such a diet contributes a degree of global imbalance, and also puts us at risk of being out of step with the season and environment where we are living—just generally out of sorts.

If people had more ideas of how to eat more vegetables in easily prepared and highly palatable ways, they would not be so quick to give up the CSA weekly box. In recognition of this reality, some CSAs have produced guidebooks, or cookbooks, for member households, to support them in understanding and using the bounty and variety that comes from the farm (see *Appendix F—Resources*).

Community Rhythm

The consumer is the weakest link in the chain of relationships established by a typical CSA. In this, the manifold pressures of the modern, harried lifestyle are the chief factor. Another factor is a general lack of understanding about what CSA is doing, and trying to do. Member households need consumer education. The most committed members tend to speak about CSA in moral, or spiritual terms. The less-committed members talk about it in economic terms, and often cite the inconveniences associated with belonging.

As observed by author and CSA farmer Michael Abelman, "CSA is a wonderful economic model, but I find that in many instances it falls short of the 'community' part of the name. Based on what I have seen, CSA projects often have trouble getting the members involved beyond the box of food. People's lives are so busy that it is tough for them to add one more thing—involvement with a CSA—to their list of activities and responsibilities."

"We need to be careful in how we present CSA to the public," he argues. "The goal is community, but that is not always realized. In the more rural areas, there seems to be more community involvement. In urban areas, there seems to be more of a simple food-financial support connection. There are some social benefits, but the basic decisions are economic."

"The pubic perception of the CSA movement needs separation between the ideal we are reaching toward, and the reality. The typical CSA garden is often a mirror of what is happening in society at large."

As there is a rhythm to the seasons, so also is there a rhythm to the growth and development of a community—something virtually all CSAs have had to learn about. A community is a group of people that has transcended its individual differences for the good of the whole. For that to happen a sacrifice is usually necessary. This is not a one-time process, but one that invariably happens over and over again.

Author Scott Peck explored the rhythm of community in a book entitled *The Different Drum: Community Making and Peace*. In the book he describes four stages through which a group typically passes before becoming a true community. People who have belonged for a time to a CSA, or to any other community group, may well recognize these stages:

- Pseudo-community. Everyone tries to fake it. Everyone wants to achieve instant community by being pleasant with one another, and avoiding all disagreements.
- Chaos. This typically arises from well-intentioned, but misguided attempts to change or convert each other. Everyone's differences are right out in the open. Members of the group are trying to make everyone "normal" and begin to fight over whose norm might prevail. At this stage it is common for the group to attack not only each other, but also the leaders. Everything can seem hopeless and unworthy. But if the members of the group remain committed, solutions begin to arise.
- Emptiness. Each member tries to empty him or herself of barriers to communication, such as expectations, prejudices, ideology, and the need to convert or control others. This emptiness can be frightening—but typically produces an appreciation and celebration of differences. This emptiness of ego, Peck writes, is the most critical and often the most difficult part of the process—but is essential.
- Trust. When the chaos has subsided and members of the group have emptied themselves of the blocks which stood in the way of their coming together, a space of trust is created. Members begin to share feelings and perceptions, not ideologies. Bonding takes place, and community happens.

Farm Festival

One expression of the yearly cycle and its festivals is presented by Eve-lyn Francis Derry in her book, *The Christian Year,* which offers a series of powerfully eloquent word pictures. Through autumn and into win-ter, she suggests, the earth inhales, drawing life forces deep within her blue-green body. The planet's surface becomes bare and still at this time, but below she is most active. In a sense, the earth is thinking deeply the thoughts which are being sent down to her as light from the planets, stars, and swirling comets that mark the limitless expanses of the sky.

When the earth exhales in spring, life returns to the surface. The vitality expands in summer and the planet's surface covers itself with visible proof of life: leaves, blossoms, buzzing insects, and rising pas-sions. In a sense, earth puts forth what she has thought in the winter in the form of summer's growing and flowering things. As the cycle of the year continues around its spiral wheel toward autumn, the earth begins again to inhale, to draw the life forces back in.

Four great festivals mark the turning points in this planetary breath-ing rhythm: two solstices, two equinoxes. At these turning points, we have opportunities to pause, moving our attention from matters of sur-vival or amusement, to the very forces which bind us with the Earth and with eternity—forces which can remind us that we are not solitary creatures living isolated lives, but rather threads in a phenomenal tapes-try of life that includes everything from the sand to the stars. Through the theories of Relativity, Quantum Mechanics, Uncertainty, and Superstrings, modern scientists have begun at last to glimpse this spiri-tual fact.

While one may appreciate intellectually the lofty-sounding concepts of rhythm, or unity, the raw power inherent in them is unavailable until it is acknowledged, called, experienced, and then consciously expressed. For this reason, for thousands of years, people have come together dur-ing the festival times to invoke spiritual energies, and then to express them in dance, song, feast, laughter, theater, and ritual.

The basic pattern of a rhythm is simple enough: activity-rest/activ-ity-rest/activity-rest. Yet there are distinctions, variations, accents, and permutations. These are punctuated by celebration at various festival times: the planting festival, the harvest festival, the thanksgiving festival, and so forth—all natural to the rhythm of the farm, all natural to the life of a healthy community.

This is something that many CSAs are rediscovering. Through farm community gatherings, members are acknowledging the real rhythm of the year, the real rhythm of the farm that provides their food, and the real need for people to unite in community.

From its inception, the Temple-Wilton Community Farm in New Hampshire has sought to establish a rhythm of festival in harmony with the farm, and the lives of the community members. The farm's dairyman, Lincoln Geiger, explains: "When people come to a CSA they are looking for different things. Some are looking for high-quality food. Some are looking for natural food. And some are looking for some kind of a community experience. The sense of what the glue is that holds the community together varies from family to family.

"To strengthen that community sense, the newsletters, the sharing of the harvest, and the festivals are essential. Together they create the substance of the farm for the members. That's where the community meets the farm, and that's where they see what we are doing.

As Lincoln explains it, the planning for their festivals arises from the Thursday Group, which is the Temple-Wilton Farm's steering committee. It meets the second Thursday of every month, and carries responsibility for the newsletter, the budget, major investments, future long-term planning, and so forth.

When it's getting close to one of four festival times that the farm observes, a member of the Thursday Group will volunteer to take on responsibility for planning and hosting it. Ultimately, perhaps 50 percent of the membership will turn out for a given festival, depending on the members' schedules and rhythm of life. Many of the farm's neighbors will also drop by.

"At the Harvest Festival in the Autumn there will be a large display of vegetables and lots of colored leaves," Lincoln says. "We decorate outside and put out the whole store for that day. People pick up their food from the display. And we also have a farm walk that day, and some music or singing. We have pumpkin and apple pie, and apple cider. Many times we'll have the children perform—and sing something for us, or enact a play.

"The May Festival is similar, except we don't display anything, since we are just then planting. People gather from 3 to 6 on a day during the first weekend of May. There's a May Pole, lots of singing and music, and refreshments. Towards the end of June we have a St. John's festival to celebrate the summer solstice. We gather round a big fire in the evening and there is singing and poetry, followed by swimming and socializing.

"Then we have a Winter Festival that we hold at the end of December, just before the New Year. It's a sharing of artistic talents—people in the farm community who have something to share will come. Some may play the piano, or this year someone played the harp.

"A community that's hung together for a period of time, and faced challenges and successes, has a life of its own," Lincoln observes. "This is true not only in the sense that the farm is a presence in the neighborhood, but it is also a shared consciousness among the membership. The festivals really bring that home. Festivals that celebrate the gifts of each season bring the spirit of a higher life down to our earth, and this higher life brings honesty, integrity, and hope."

Essay 3

Large-Scale Community Farms

Most CSAs are modest in scale, involving perhaps twenty to 100 member households, and five to ten acres of land. But some have expanded far beyond these parameters. In the United States, the CSA concept first took hold on the East Coast, then swiftly spread to the Midwest, and the far West. The experience of the West is different from both the East and the Midwest, and worthy of its own discussion, for it is home to more large-scale CSAs. Different dynamics come into play.

Jered Lawson has had a good vantage to observe trends in the nearly 100 CSAs that have formed in California by 1997. He first learned of CSA from the Live Power Community Farm in 1989, then went on to complete a national study on the CSA movement in 1991 while a student at the University of California, Santa Cruz. Jered was instrumental in the development of the Homeless Garden Project CSA, and a CSA for the Center for Agroecology and Sustainable Food Systems at the UCSC Farm and Garden. From 1994 through 1996 he initiated and then served as Coordinator of CSA West, a program of CAFF—California Alliance of Family Farms, an organization that supports West Coast CSAs, predominantly in California (see *Appendix F—Resources*).

Jered notes distinctions in the growth and development of CSA on the West Coast as compared with the early CSA development on the East Coast. "The original CSA idea was that a group within a community—an already organized group or just a number of people in a given geographic area—would commit to support a farm for the entire year. They would sign a commitment form that says 'regardless of large or small harvests, through thick or thin, we will see that your farm's budget is covered and that you are able to survive, at least through this season.'

"When I first learned about CSA, I was very excited about this concept—the whole farm or garden being completely supported by a given number of people so that the work the farmers are doing can be assured and sustained through the season, and hopefully on into the future, season after season.

"In some ways though, after a time, I became a bit disillusioned by the reality that most of the California farms initiating CSA were not doing it in the same way, with the same intention. They weren't looking to cover the farm's whole budget, or even a large percentage of it through CSA.

"CSA was arising as more of an add-on concept for forty to eighty to 100-acre organic farms. They already had a large part of their production and distribution geared toward the wholesale market, toward farmer's markets, or toward other forms of direct-to-consumer marketing. CSA became just another part of their overall operation. The add-on approach to CSA in some ways limited the ability of community members to see themselves as part of the farm, and to wrap themselves around the whole farm in the way that they could in the East Coast, and thereby to be a part of the farm in a very fundamental way. The community's involvement was limited by this adaptation of the CSA model.

"It was a bit disconcerting to me at first. I wondered, does that mean that the concept will be weakened, and that the community involvement of the CSA—the social dimension—will be lessened and perhaps deteriorate over time? I was concerned that then CSA becomes 'just about vegetables,' and not about the broader and deeper possibilities expressed in the original CSA concept.

"But over time, I came to see this differently. Just looking at how organic farming has evolved in California—the demographics, the scale, and the complex pressures on a lot of farms to maintain a certain level of growth— to do this, the farms almost had to grow in this way. So, while the CSA becomes only a small percentage of their overall farm activity, they have through CSA incorporated into their farm a greater capacity for long-term sustainability.

"Even in a hybrid CSA, or a CSA that may not have the whole farm budget as its centerpiece, there's still the fundamental desire to see the farm survive. The consumer members of the farm do come to understand that they are giving economic contributions toward that survival, and the farmers themselves help set the terms of that contribution. Therefore, the consumers can say 'we feel comfortable that we are meeting the farm's assessment of what it needs to survive,' even though it is different from the original CSA concept.

"Down the line there could be potential challenges with this partial or hybrid CSA approach. For example, if wholesale markets and so forth become threatened and therefore threaten the economic stability of the farm, then the possibility of the CSA community maintaining the overall viability of the farm could also be threatened. And yet, at the same time, if wholesale markets were to fall through somehow, for some reason, the farm would have the CSA model in place, and also have familiarity with it. They could then shift emphasis to expanding the CSA dimension of their farm as a survival strategy.

"One question that comes up frequently is, what is the appropriate, or workable ratio for a CSA? When you look at the scale which is appropriate to the physical capacity of the farm, and the number of farmers to the number of people eating the food, what's the right ratio? What is the level at which the relationship can be sustained in a fashion that is mutually beneficial for the farm, the farmers, and the community? Finding that ideal ratio is a key challenge for every CSA. I've heard many people suggest that it takes three to five years to get to the point where production, distribution, community needs, and farmer needs are all balanced.

"If a 100-acre farm, which is highly mechanized because of that scale of production, is going to expand to a full CSA—with all the production going to members in boxes of food—say in an urban area—then there may be 2,000 families supporting that farm. Is that ratio too large for a true relationship to form? Or, through the newsletter and community events on the farm, can there still be enough contact to keep a core understanding of the importance of the economic and social relationship with the farm, and still retain enough of the benefit to all parties that it is worth sticking with it over the long haul? If you are one person among 2,000 how connected do you feel?

"At what level do people still identify with, and still feel a part of the core structure of the farm? What is the minimal level of relationship, whereby the farm is still strong? At what level does it become difficult, or even impossible, to keep that feeling alive?

"We are seeing more larger-scale farms here on the West Coast. The largest CSA in California right now (1996) has 700 members—The Be Wise Ranch. There are a couple of other large-scale farms, Full Belly Farm, for example, is approaching 500 members. And we've recently had a call from a farm manager who is thinking about establishing a 2,000 household CSA—and making deliveries far from the farm to households in a urban areas, potentially throughout the US.

"In education, a lower ratio of students to teachers means more meaningful education for each student. When you pass a certain number of students for each teacher, the education suffers in quality. Kids learn less. So, where is the threshold of meaningful relationship in a CSA? It is talked about generally, but I know of no definitive study. You have to look at the CSAs that have endured the longest to begin to get a sense of the appropriate scale.

"Some CSAs are at a scale that employs a large number of migrant laborers. It's not that employing migrant labor is wrong, it's simply a broader image of the farm and its scale than what CSA was originally associated with. It's become more than just the farm family with apprentices. So we must ask, how are the people a part of the life of the farm? Are they integrated into the CSA community? Or are they seen simply as a commodity, something to be bought for their ability to get the job done? Are they included in the communication, the newsletters? Do they partake in decision-making, planning, visioning? Do they have dreams of their own, and how are those shared in the community and addressed? What about their families? Do they, too, reap the harvest, or only wages? Is there food being grown that they enjoy to eat? And what about the cultural celebrations, like Spring planting parties, and summer solstice potlucks? Are they included? I've been to a great harvest festival at Full Belly Farm, where their Latino employees are very much present, with music, dancing, food, and mingling with CSA members and other visitors. This is exciting.

"I remember talking with Robyn Van En, the founder of Community Supported Agriculture of North America, Inc. (CSANA). She echoed the sentiment that we found in a survey we did back in 1992—that there are millions of people who need good food and the kind of connection to nature and community that comes through CSA. So larger CSAs are needed. But, the larger it gets, the greater the tendency to lose the personal relationships that make CSA a meaningful alternative to the anonymous and often suspicious supermarket.

"What are the indicators of health of a CSA farm? It seems the main focal point should be the relationship between the farmer and the community, and the health of that relationship. Because it is the relationship that will survive, and keep things together on a CSA, when production varies, as is typical on any farm. Farms have both good and bad years. It is the survival, then, of the farm as a community that we are striving for. The health of that relationship should be what we assess, and how do we factor in distance from the farm?

"Larger CSAs are continuing to spring up. Over the long-haul there will be an opportunity to observe how they function and prosper. This development is perhaps inevitable, especially considering the expansion-ist mind-set of the American economy in general. The driving force in agricultural change has been, and will likely continue to be, that the people in power around the agricultural sectors of the economy will want to continue to push toward centralization and globalization of farm commodities. In contrast, CSA is steadily creating a new con-stituency of active citizens who are trying to re-democratize the food system through the CSA partnership and other forms of direct farmer-to-consumer relationship.

"Here on the West Coast, where larger-scale organic farms exist, the trend is for people to come on board the CSA movement through these larger-scale operations. One could say that this is increasing the central-ization of CSA food production. There is a trend toward that same kind of capitalistic impulse to grow bigger, to have labor still seen as input, and to keep the cost of that labor low so that the farmer can make a higher profit. The farmer may not be somebody who is actually working the field, planting seeds and pulling weeds, but maybe somebody who is sitting behind a desk and managing the whole operation, or maybe even off somewhere else, having hired someone else to manage it. As I see it, there's a possibility that the CSA movement in California could be co-opted by these larger global forces.

"Yet we can still uphold a vision of more and more small-scale CSAs dotting the landscapes, both rural and urban, filling every neighbor-hood, every community, with personal and meaningful connections to their farmers and farms. Building a new food system that's about rela-tionships, not price per pound, is something that may involve greater challenges for larger-scale CSAs, but then this is surely a global trend that we are up against. As long as large-scale CSAs can evolve with, and not eclipse small-scale CSAs, then in each the CSA membership will have the opportunity to keep moving toward greater conscious support of their farm's, and society's, general economic, social, ecological, and cultural health. Then we will be headed in the right direction."

Essay 4

Community Farm Coalitions
Marcia Ostrom

Cooperation among community farms is central to transforming regional food systems and to the success of the CSA movement as a whole. Described by some as the "next level" of CSA, building connections between farms creates a new dimension for furthering the ideals and practice of CSA. Collaboration at the farm-to-farm level enables community supported farms to develop and exchange practical knowledge and information with each other, coordinate public outreach, pool material resources, and develop support systems for new or struggling farmers. This is also an important arena in which various perspectives, values, ideas and philosophies about CSA can be presented and negotiated, resulting in the creation of larger shared visions and new ideas for bringing about social change.

It is imperative for the long-term viability of the CSA movement that these connections among farms be nurtured because when community farms view themselves in isolation they can begin to compete with each other for members and resources. Besides losing out on the kinds of support described above, there is serious potential for undermining the basic values and economic foundations of the movement. Only by cultivating coalitions with one another and with other types of farms and community organizations, can community supported farms begin to develop the kinds of support structures which offer a foundation for creating and sustaining a new form of agriculture.

In Wisconsin and Minnesota mutually supportive relationships have developed among CSA farms through formal and informal networks. These coalitions have played a pivotal role not only in establishing CSA in the Upper Midwest, but also in maintaining the health of the movement. Two associations have emerged as key centers of such collaborative activities. Formed in 1992, the Madison Area Community Supported Agriculture Coalition (MACSAC) includes sixteen community supported

Marcia Ostrom is a Ph.D. candidate in Environmental Studies at the University of Wisconsin.

farms, as well as community activists, aspiring CSA farmers, and other supporters. The Minnesota-Western Wisconsin Community Farm Association (MWCFA) was founded in 1993 and consists of twenty-six CSA farms.

Just as each CSA farm has developed its own distinctive organizational character, these associations of farms and activists are continually evolving and remaking themselves. While both groups evidence a high degree of cooperative spirit and operate by consensus decision-making, their organizational forms, goals, activities and roles have evolved differently.

The Madison Area CSA Coalition

MACSAC originated with a small group of activists who began gathering together in their living rooms in the fall of 1992. This group questioned why their town, situated amidst a fertile agricultural belt, should be importing the majority of its food from great distances. CSA appeared to offer an innovative strategy for addressing many of the problems they saw with such a food system.

Over potluck meals, this group painstakingly worked to establish consensus on what they felt were the priorities and principles of CSA: to embody the ecological principles of sustainable agriculture, to build community, to provide a way to share the responsibilities, risks, and rewards of farming, and to provide an educational opportunity for both farmers and urban consumers on many issues including nutrition. They felt strongly that CSA should be accessible to all people, regardless of their income. As a group spirit began to take shape over the course of these discussions and potluck dinners, so did an underlying philosophy and commitment to take action. By the third meeting they had generated a timeline getting one or more CSA farms started by the next growing season.

An important topic of discussion was whether to launch a single farm (there was already a farmer in the group) or multiple farms. As an initial step, group members surveyed their friends and relatives to assess interest and levels of commitment to the idea of CSA. They also began gathering names and information about potential growers in the area. Once favorable feedback was received from the consumer surveys, existing vegetable growers were polled for their openness to the idea of CSA. The response from area growers indicated a widespread interest in CSA. A producer's meeting was arranged for January. Dan Guenthner from Common Harvest Farm, a community farm in Minneapolis, was invited to make a presentation and field questions about CSA farming.

Enthusiasm for CSA among the assembled farmers ran high. The group ended up deciding to help eight different CSA farms get started that first year.

The next step was to figure out how to reach the broader Madison public. By this time, the group had come up with the public name of Madison Area CSA Coalition (MACSAC). Brainstorming sessions on the topic of public outreach resulted in the idea of holding a CSA farm fair in early March. An open-house format would allow MACSAC farmers to set up tables and distribute information about CSA and about their farms and potential members could meet directly with farmers.

In addition to making plans for the farm fair, campus and community groups were informed through a series of talks. A few people began working diligently to cultivate media support. Farm directories and informational brochures were developed for distribution through supportive local organizations. Growers determined from the outset that share prices should not be listed in the farm directory to discourage bargain shopping among consumers and competition between farmers. The idea was to encourage consumers to contact individual farmers and talk with them firsthand about their CSA operations.

With the help of a superb location and friendly media coverage, the MACSAC Open House drew more than 300 people, a far larger crowd than imagined, initiating an ongoing annual tradition. The combination of the open house, good media publicity, informational talks around town, and the farmers own efforts to mobilize communities of friends and acquaintances helped all the farms to gradually realize their membership goals. The eight farms ranged in size from four to fifty-four shares, putting the total number of Madison households participating at somewhere around 260.

The 1993 season was off to a promising start until it began to rain. Farms experienced flooding ranging from mild to severe, but nearly everyone had lower and later yields than expected. One farm lost an estimated 80 percent of the crops they had planted. The farmers tried to help each other as best they could by sharing or swapping any surplus produce. One farmer felt so desperate that he purchased wholesale organic produce to distribute to his farm members. He said that if he hadn't been a CSA farmer it would have been better because he could have just quit for the season.

The circumstances brought on by the weather were not all bad, however. One farm credited their desperate pleas for volunteer help with engendering a deep spirit of cooperation and commitment to the farm on the part of their members. They discovered that "people really want to help if they feel needed." This farm was able to summon twelve volunteers on a moment's notice to get their transplanting done during a window of dry weather. "It was really uplifting," the farmer commented, "like having a big party out there in the field." This farm credits the challenge of the adverse weather conditions with really bringing the farm together and teaching participants the value of CSA.

At the first post-season MACSAC meeting, the highs and lows of the 1993 season were the subject of intense discussion. With the exception of one farmer who decided that his operation was really better suited to growing for market, all the farms committed to continuing for another season. Since that first year, the number of CSA farms and the size of the existing farms in Madison have expanded each year. From serving around 260 households in 1993, four years later sixteen MACSAC farms are feeding an estimated 1300 households. The farms range in size from very small (four shares) to large (300 shares).

MASCAC Goals

MASCAC was originally founded with the mission of educating consumers and farmers in the Madison area about community supported agriculture. While a mission statement has never been formalized, the working draft declares a commitment to promoting a healthy, local food supply, supporting small-scale farms, building community and protecting the environment. Over time, the goals of MACSAC have gradually shifted from a focus on getting CSA established in Madison to maintaining and supporting existing CSA farms.

Consumer Outreach

MACSAC organizes a variety of activities targeted to consumers. Each spring they host a CSA Open House at the city botanical gardens. In a celebratory atmosphere, this event, which sometimes draws up to 500 people, offers keynote speakers, activities for children, educational slideshows and opportunities to meet personally with farmers. Additionally, MACSAC continues to maintain an updated directory of all the CSA farms serving the Madison area. They also supply speakers and displays on CSA to local community groups and are involved in projects which seek to provide fresh, organic produce to people with low incomes.

Recently, MACSAC created *From Asparagus to Zucchini,* an impressive "food book" targeted towards CSA members and other consumers.

With an alphabetical recipe and information section for each local CSA vegetable, this book (see *Appendix F—Resources*) details ways to use and preserve farm-fresh seasonal produce. It was designed to address the concerns expressed by consumers as they struggle to overcome "supermarket withdrawal," seasonal vegetable fluctuations, onslaughts of unknown vegetables, root crops tedium, and other typical consumer challenges. Mixed in with the recipes are educational tidbits about CSA and the larger food and agricultural system and descriptions of each local farm. The book has been in demand by people across the country and may prove to be MACSAC's most successful fund-raiser.

Farmers have found that working together to do public outreach benefits everyone. One farmer stated that "having been in a position where I was the only one who was doing it, I know for a fact that it's a lot easier to promote the concept of CSA as part of an organization than as an individual." He went on to explain that as an individual, consumer outreach felt very self-promoting. "It always seems like you're just trying to sell your own stuff." In contrast, as part of a larger group, the organization can present the idea and let the consumers choose how they want to get involved without any pressure. Farms feel that publishing a list of farms makes it much easier for the general population to get in touch with them and leads to increased phone inquiries overall. As one farmer summed it up: "you always have more clout, more power, and more credibility as a group than you do as an individual."

Building Community Among Farmers

MACSAC farmers get together on both a formal and an informal basis. Meetings are organized around potluck meals to encourage a sense of community and provide space for social interaction. Farmers say that developing informal relationships with each other is one of the most important apects of MACSAC. Farmers report such cooperative efforts as doing seed orders together, trying out each others' equipment, sharing a dump truck for compost, placing bulk orders together, sharing surplus produce, getting advice over the telephone, and so forth.

One farmer commented on this community of farmers as follows: "I think a lot of the real stuff happens outside our more formal bureaucratic meetings. A neighboring farmer and I are really getting to know each other now. We're talking about equipment sharing and I watered his stuff while he was out of town. . . . That kind of little day-to-day stuff is important. I see that happening throughout all the different members of MACSAC."

There is potential for cooperation to deepen still further in the future. Ideas that have come up include planting extra crops for insurance in case someone has a shortage, having some farms grow enough of certain difficult crops to supply other farms, and also creating a health insurance pool to bring down rates.

Another important aspect of working together is the development of a sense of common purpose and a shared vision of creating change in the current agricultural system. This sense of common goals can provide an emotional lift during difficult times. One person observes that "it's very important that we're all doing this together and putting this new idea out there with integrity and high quality." It helps her to know that "we're all doing this for more than just our own individual livelihoods, but really investing in the bigger picture, moving towards some new agricultural ways."

Knowledge Exchange

Like other forms of ecologically-based farming, community supported agriculture is knowledge intensive, but even more so since it also has a social dimension. Because knowledge of regenerative farming practices and community organizing are not readily available through other channels, informal knowledge transfer between CSA operations takes on a high level of significance. Recognizing this need, MACSAC farmers have organized meetings around a variety of practical topics, such as equipment, various aspects of vegetable production, seed ordering, building core groups, setting share prices, and so forth. Farmers really come to count on the practical knowledge they gain from each other.

For the past two summers MACSAC has organized farm tours. This is a great opportunity to learn about the things that really work for other people. Conferences have provided another helpful learning tool. For the Upper Midwest CSA Conference, held for the past three years, farmers have planned workshops around topics of concern to them. Farmers and others viewed as experts in these areas are then invited to make presentations and lead discussions. One farmer explains that "we share a lot of interests and a lot of common problems, day-to-day problems. When we're able to communicate about those concerns with each other, we can generate new ideas and solutions to problems."

Organizational Form

Each fall after the root crops have been harvested and stored, MACSAC members come together to discuss the past season, and to rethink and debate anew the optimal structure for their organization. Since the days of gathering in living rooms, the form of the organization has taken several different turns. In the beginning when the group was small and composed primarily of non-farmers, MACSAC met frequently, even during the growing season. As the group expanded and came to be increasingly composed of farmers, time and distance made coming together more difficult. Over time, many MACSAC tasks have been decentralized and delegated to committees. Full meetings are currently held twice a year: at the end of the growing season and in early spring. A central coordinating committee has been empowered to make decisions that cannot wait to go before the full group. Other committees address issues of outreach and education, outreach to low-income communities, the spring event, and grower activities. Individuals volunteer for roles such as secretary, treasurer, chair, and media liaison.

While MACSAC does not have any paid staff, a few jobs such as secretary and media liaison have proven so onerous that the people performing these roles are offered monetary compensation for their time. Also, the Wisconsin Rural Development Center (WRDC) in Mt. Horeb has played an important supportive role from the outset. Besides funding a staff member to play an active role in founding MACSAC, WRDC has provided important financial services. Since MACSAC is not formally organized as a 501 (c) 3 non-profit, it has relied on WRDC to be its fiscal agent. Revenue is raised through dues (farmers pay a fee based on their number of shares), small state and local grant programs, and, more recently, through sales of the food book. Over time, WRDC has also taken on the role of primary phone contact for MACSAC and general CSA clearinghouse (see *Appendix F—Resources*). The number of phone inquiries about CSA have far exceeded the capacity of MACSAC volunteers.

The overwhelming interest in CSA on the part of area farmers raised a number of difficult questions early on about who could join MACSAC. The group felt that it was really important that CSA farms made every effort to ensure that farm members were having a good experience. How could MACSAC ensure high quality, ecologically sound production? What if someone who isn't organic wants to be a CSA grower? Is just being organic going far enough? What if someone really doesn't have the knowledge, experience, or the practical resources to be a good grower? What exactly constitutes a CSA farm? What about people who follow more of a subscription farming model? What about specialty growers?

In the end, while never a comfortable role, MACSAC has been able to retain some degree of control over who to support as CSA growers through deciding which farms to include in the farm directory and invite to the spring open house. When new farms ask to join MACSAC they are provided with a questionnaire about their growing practices and plans for organizing a CSA community. To be included on the list for the upcoming growing season and to have a table at the open house, farms need to have experience producing diverse vegetable crops, have knowledge of alternative pest control and fertility practices, have access to land, have a plan for building a CSA community, and be prepared to pay dues and volunteer with MACSAC projects. Farmers lacking experience or knowledge are encouraged to intern or apprentice with already existing farms. Specialty growers who approach MACSAC are encouraged to develop cooperative relationships with the existing CSA growers, and to develop their own direct marketing networks.

MACSAC Challenges

One downside of having such a strong network of farms in the Madison area may be less loyalty to a particular farm on the part of members. If members dislike a particular aspect of belonging to one of the farms, they may be tempted to simply change farms rather than stick it out and make an effort to improve the situation at their current farm. In an extreme example, if a farm loses its farmer, ideally, that farm community could get organized and find another farmer as has happened on occasion with Eastern U.S. and European CSA farms. But in the Madison area, the response to losing a farmer has been for individuals to disperse and join other farms. Similarly, it has been difficult to get members involved in issues of land acquisition. There doesn't seem to be a critical need to sacrifice in order to pull together and preserve a particular CSA project when there are so many others nearby to choose from. With so many farms to choose from, consumers might have less stake in making sure that the economic and social foundations of their farm are secure.

A related problem is one of weak organizational structures on the part of individual farms. At this point, CSA in Madison is still largely farmer-driven. It has been difficult to foster a deep sense of involvement on the part of consumers and many of the MACSAC farms struggle to build strong and dynamic core groups. The tendency is for the farmers to do everything without asking or relying on members to carry out any significant portions of the organizational work of planning, harvesting, distribution, communication with other members, and so forth. As a result, farmers often feel overworked and exhausted. Part of this problem may lie in the fact that farms initially relied on MACSAC to perform many functions of a traditional core group rather than developing the capacities of their own farms.

As of early 1997, MACSAC still has not succeeded in involving low-income people in CSA to a very great extent. Individual farms have created opportunities for worker shares and subsidized shares. In some cases these options have been utilized and really appreciated, however there is still a long way to go in identifying and systematically meeting the particular needs and situations of people in lower income communities. MACSAC has just received a grant from the "Wisconsin Food Systems Partnership" funded by Kellogg to further address this priority.

Finally, a potential problem that people in MACSAC worry about is size. MACSAC has grown from a small cozy group of friends meeting in living rooms to the kind of group you have to reserve meeting halls for. Some people are wondering how big MACSAC can get before it loses its friendly character and ability to perform its mission. One farmer said "I worry that we're going to get to the point where we're just shooting ourselves in the foot by being so cooperative and having this open policy for anyone and helping everyone to get going." The fear is that a town the size of Madison may not have infinite numbers of potential members. Are all these farms going to be able to make it? With such a small population base, it will be imperative to focus on member retention and reaching out to new groups of consumers.

The Minnesota/Western Wisconsin Community Farm Association

Verna Kragnes of Philadelphia Community Farm traces the roots of the Minnesota-Western Wisconsin Community Farm Association (MWCFA) to a public radio interview she gave in the summer of 1991. At the end of her presentation on CSA, the director of the Minnesota Food Association (MFA), spoke for a few minutes and offered his organization as a contact point for obtaining further information on CSA. The number of calls generated by this twenty-minute radio spot far exceeded any publicity efforts in MFA's ten year history. Ken Taylor, the director, was so impressed with the level of local interest in CSA that he invited Verna Kragnes and Dan Guenthner of Common Harvest Farm, the pioneering CSA growers in the Minneapolis/St. Paul region, and the Land Stewardship Project (LSP), another local sustainable agriculture organization, to hold a meeting on CSA. Out of this meeting arose the idea of inviting Trauger Groh to speak in Minnesota in January of 1992.

Trauger's first visit to the Twin Cities area marked a critical juncture in the seriousness with which local community leaders viewed CSA. An afternoon question and answer session for farmers overflowed the designated space, leaving standing room only for latecomers. Later that evening 250 people jammed an auditorium at Hamline University to hear Trauger speak persuasively about the need for a new kind of farming. He eloquently described the death of family farming as we know it, and the rebirth of new farms based on the support of communities. The energy in the auditorium was charged. Discussion groups following the talk triggered a spirited dialogue between Trauger Groh and members of the audience. The organizers had expected some interest in the talk, but the intensity of the response left them shaking their heads.

To meet the anticipated flurry of informational requests, Verna and Dan had planned a series of events following Trauger Groh's talk. They held a day-long "getting started" workshop for prospective farmers and then a series of three more informational meetings for consumers. Rather than stemming requests for information, the workshops generated still more calls to Dan and Verna. In the face of such sustained enthusiasm, a "CSA Task Force" was formed consisting of the two CSA farmers, Verna and Dan, and representatives of LSP and MFA.

The newly formed "CSA Task Force" decided to sponsor a study circle in hopes of bringing local understanding of CSA to a deeper level. The goals of the study circle were described as: creating a shared vision of community supported agriculture in the Upper Midwest, evaluating how community supported agriculture can be put into practice, and recommending guidelines for the future development of community supported agriculture. Monthly meetings were scheduled for the fall of 1992 and the spring of 1993. Forty people applied for twenty available slots in the study circle. Applicants were selected with an eye towards creating a diversity of urban, rural, gender, farmer, non-farmer and professional representation. Each discussion session covered a topic, ranging from the basis of ecological thinking to the global food and agriculture system to the idea of a foodshed. The group studied both the Japanese and the biodynamic models of CSA, as well as the history of food production in their area. The group generated a report summarizing and tying together what they had learned, and the report was widely circulated.

The study circle planted important seeds. Many of its members went on to join the next generation of CSA farms which were springing up around the Twin Cities, forming the basis for strong core groups on these fledgling farms. Other members of the study circle went back to the organizations they were members of and got them involved with CSA. Finally, the report from the study circle fueled efforts to obtain grants for further educational activities around CSA.

At the close of the study circle, the Task Force called the old and new CSA farms to a meeting. At this meeting the Task Force officially disbanded itself and asked the farmers to determine what would happen next. Verna explained that "we felt it was time for the group to have broader representation." Thus, in the spring of 1993 an association of CSA farms and other interested individuals and organizations was initiated. They met at a church in Minneapolis and then at the log house at Philadelphia Community Farm. A directory of farms was developed with the help of the Minnesota Food Association and the Land Stewardship Project, which also became local contact points for CSA information. This stage saw a gradual transferal of leadership from the members of the task force to the thirteen member farms.

The next fall the new association decided to hold a cultural event in conjunction with a CSA workshop; they called it a "Celebration of Farms." On a Friday night, they had a children's storyteller and on Saturday, a workshop led by Ian Robb (see *Example 3—The Brookfield Farm*). From a mailing list of thiry-five farms, over 105 farmers and consumers showed up and the workshop had to be relocated to a nearby Waldorf School.

Subsequent meetings of the new association that autumn resulted in the adoption of the name Minnesota/Western Wisconsin CSA Association and the forging of a mission statement. As of early 1997 the group is still meeting and there are twenty-six member farms. The association is composed primarily of farmers, farm family members and apprentices, although sometimes a representative from LSP or other interested individuals attend meetings.

MWCFA Goals

The association was originally founded to support new farms and farmers, to share farming skills, to facilitate farmer information exchange, and to organize joint efforts at publicity. In a discussion about the mission of their group, association members named important functions as helping farmers get to know one another, providing inspiration for each other, and building community among farmers. The official mission statement approved by this group in the fall of 1993 reads:

"The Minnesota/Western Wisconsin Community Farm Association

is an association of individuals, farms, and other organizations working together to further the establishment of CSA in the Twin Cities and Western Wisconsin in order to revitalize community and reconnect people to nature; ensure healthy food and a reformed sustainable food system; and protect and restore agricultural land, ground water, and nature resources in perpetuity."

Information Exchange, Economic Cooperation, And Community Building

In the past, information exchange among MWCFA farmers has been facilitated through a winter series of grower workshops on technical topics. More recently, growers have gotten together for what they call "Vegetables from A to Z." Just before seed ordering time, growers sit down and go through the vegetables one by one in alphabetical order and share what they know with each other. This is described as "very effective because it gives you a chance to learn and do some visioning before seed ordering time." The annual Upper Midwest CSA conferences, organized jointly by MWCFA and MACSAC, have been important forums for sharing and learning as well. MWCFA farmers have played central roles in coordinating and setting up workshop agenda and keynote speeches for these conferences in accordance with local needs.

Philadelphia Community Farm has been active in spearheading grants to help with developing and disseminating knowledge about CSA. They have received money from the USDA through its SARE (Sustainable Agriculture Research and Education) program and from the state of Wisconsin to carry out research projects, conferences, field days, and study circles. They plan to print up an informational booklet for farmers documenting what they have learned.

Apprenticeship programs are currently being developed. A committee is working out an apprentice exchange program based on a model from the Northeast. Area apprentices will get to work on different farms during the season. Farms will also host field days for all the apprentices which will be organized by the apprentices themselves. There has been some talk about developing shared housing for the apprentices in the future to enhance their experience and sense of community. Every effort is being made to help make apprentice work a pleasant and educational experience. Besides learning for themselves, by traveling from farm to farm, the apprentices facilitate information exchange among farmers.

A number of CSA farms have gotten together to form a producer's pool in order to generate additional income. They have contracted with a regional grocery store chain to supply organic produce during the 1997 growing season. LSP is providing a part-time organizer to help get things set up and running smoothly. The farmers involved plan to plant beyond their CSA needs to have extra produce to retail.

The most important aspect in building a spirit of cooperation is just getting farmers together in the same room. One MWCFA member explains that getting farmers together over good food to discuss problems that they have in common is a sure recipe for creating a sense of community. He feels that to help farmers get comfortable with each other, it is best to bring them together initially for practical purposes such as the grower discussions. The annual 4th of July picnic provides an opportunity for farmers to gather at one farm during mid-season to socialize and discuss practical issues. These events are surprisingly well-attended given how busy the farms are at this time of year.

Public Outreach

MWCFA has sponsored a variety of events designed to provide information about CSA to the general public. They have used talks by well-known speakers such as Trauger Groh and Ian Robb to appeal to a broad-based audience. They have followed up such events with opportunities for hands-on workshops and discussions. Study circles, held in 1992 and 1995, offered an important opportunity for interested farmers and community members to reflect on the problems encountered in applying the concepts of CSA.

Cooking classes, targeted towards farm members were held during the winters of 1994–95 and 1995–96 in the Twin Cities. Expert chefs were brought in to teach farm members creative ways of utilizing their CSA vegetable shares. These classes took cooking with CSA vegetables to new culinary heights and were very popular with farm members.

In cooperation with The Land Stewardship Project (LSP) and Minnesota Food Association (MFA), a yearly directory of farms is printed. Each spring when farms are looking for members, LSP helps to arrange for media publicity focused on CSA and area community farms. LSP and MFA serve as contact points for community members who want to learn more about CSA or how to join a farm.

Organizational Form

At its conception some group members argued for a purposefully loose organizational form which would permit people to contribute and participate as they are able. For the time being these forces seem to have won out. One member remarked that "in many ways, our group is so loosely organized and affiliated that it's remarkable it hangs together." Another likened their organization to a plant growing, saying that "the growth keeps pace with what needs to happen."

The MWCFA generally has bimonthly farmer meetings throughout the winter, and two larger meetings designed to include consumers, organizational representatives, and other interested parties. All meetings are organized and run by the farmers, with different individuals taking turns facilitating and getting everyone together. At each meeting someone volunteers to put together the next one. Decision-making is described as "consensus with a lot of strong, independent voices." Because there is such a high level of trust and caring, a strong sense of group cohesion seems to have evolved. People come with a lot of different opinions and experiences, but one person notes that "there's an unspoken code of respecting the feelings in the room."

Subgroups have formed by region. In the Stillwater area, a group for first-year farmers met every week for a while. Another subgroup has grown up around the little town of Prairie Farm where five CSA farms are located. This group holds Friday lunches at different farms during the season. They discuss growing issues and give each other support when the going gets rough. This group cooperates on apprentice programs and has discussed forming a capital fund from a possible economic development grant. They do a small amount of equipment sharing and seed ordering together and they are thinking about investing in some fertility equipment in the future. Other regional clusters of farms which tend to work together include Osceola and Northfield.

From its inception, MWCFA has had a close relationship with the LSP and the MFA. Indeed, LSP and MFA deserve a lot of credit for the enthusiastic response to CSA in the Twin Cities area. Much of the strength of the CSA movement in this region can be attributed to the groundwork laid through more than a decade of educating consumers and farmers about the goals of sustainable agriculture. These organizations helped to create an urban populace which was highly educated and concerned about the underlying problems of the existing food and agriculture system. More directly, LSP and MFA were active in launching MWCFA as described and they continue to fill important support roles, such as maintaining the farm directory, defraying the cost of mailings, organizing media publicity, sponsoring conferences, and serving as an informational clearinghouse on CSA. They have also contributed staff time for attending meetings and organizing public outreach activities.

Recently, LSP has offered a staff member to help with organizing the producer's pool. One MWCFA member observed that the help of supportive organizations such as LSP and MFA is critical in the beginning stages of building farm networks because they can bring together farmers who don't know each other and get them started talking.

Challenges

While some feel that the MWCFA suffers from being too loosely organized and would like to see the organization develop more structure and direction, others seem satisfied with the form it has taken. They feel that the organization is fulfilling its central purpose of bringing farmers together and will evolve in accordance with the needs of its members. Certainly the group has proven capable of pulling together whenever there is an important task or activity to accomplish such as a conference or public event. It has also proven resilient to the challenges brought on by long driving distances and uncooperative winter weather.

Perhaps because the Land Stewardship Project, the Minnesota Food Association, and Philadelphia Farm have shouldered so much responsibility for fund raising and public outreach , MWCFA has never really had to develop a sophisticated organizational structure. As long as other organizations are willing to handle such tasks as publishing the farm list, media publicity, finances and grant writing, and fielding phone calls, MWCFA seems to be able to subsist on a simple organizational structure. This may even be a strength if it enables them to focus their energies elsewhere. At least one member, however, would like to see the organization do some reexamination of its roots to make sure that they are on track with their core mission and underlying values. Another member is concerned that exciting new projects like the producer's pool can potentially sidetrack the group from its main purpose.

Cornerstone for the Future of CSA

CSA is a movement based on values and ideals. Each group started with a deliberate effort to create a shared vision of what CSA meant and how it could be practiced. Serious discussion took place around the different existing models of CSA. Shared values were set out on the table. In both groups, deliberate steps were taken to spread ownership of the organization from a few core organizers to the larger group membership made up of area CSA farmers. Careful attention was paid to building a sense of trust and community among group members through such social activities as potlucks and celebrations. Both groups have developed strong organizational blueprints, participatory decision-making structures and dynamic communities. One of the most important functions of each group has been to foster a cooperative spirit and sense of community among growers.

Consumer representation in such coalitions of community farms appears difficult to build. Each group has wondered at some point whether separate forums should be organized for consumer and farmer components of the movement, but in the end the consumer motivation to organize at an extra-farm level appears to be lacking. Even where conferences and workshops have been designed to address issues raised by farm members, it has been difficult to enlist their participation. While MACSAC does have a core of committed non-farmers who are central to the organization, few of these members actually come from any of the area CSA farms. Consumers appear to play a less significant role in MWCFA.

The important role of supportive organizations cannot be overlooked. They are especially important in the formative stages of organizing farm associations. Charismatic movement leaders can also play an important part in getting the ball rolling. Visits to the Twin Cities by Trauger Groh and Ian Robb helped to establish credibility and initiate provocative dialogue in the early stages. Early on farmers such as Dan Guenthner and Verna Kragnes played important roles as inspirational speakers and sources of information for other farmers. These first CSA farms set an important precedent for cooperation and sharing which planted the seeds for the development of a number of different regional networks later on.

In many ways such farm associations are the carriers and creators of the CSA movement ideals in the Upper Midwest. They are also the source of many new ideas and centers of innovation. These groups of farmers and community activists are the driving forces behind creating and maintaining healthy community supported farms. Farmers working together through networks create the spaces and structures in which visionary thinking can occur, and conscious strategic efforts can be made to strengthen the movement through forging deeper understandings on the part of farmers, consumers, and the general public. It is here that problems in the larger food system, in the region, and on individual farms are identified and solutions are constructed. Grappling with these issues provides the cornerstone for the future of CSA.

Examples of
Community Supported Farms

Steven McFadden

Example 1

The Temple-Wilton Community Farm

The Farm 1990*

Lean and thoughtful, his arms stretched long by the weight of the milk pails he carries, Lincoln Geiger makes his way through the barn at Echo Farm. His labor is hard and his body shows it. But he has his satisfactions. He has found a way to work full time as a New Hampshire farmer, and to do it in harmony with the earth. Nowadays, when most food arrives home wrapped in cellophane from parts unknown, Lincoln Geiger is a community treasure.

The good fortune of his circumstances can be traced directly to the tenacity with which he and his colleagues have held to their vision, as well as to the support of about sixty-three independent families in Temple and Wilton. Together they are creating a community farm that may well be a model for revitalizing the way we grow food. It may also be a model for reconnecting the region's farms to the communities where they are located.

The name of the enterprise, which in 1990 included not only Echo Farm but three others as well, is the Temple-Wilton Community Farm. Though they have been in existence only since 1986, they have already brought a tremendous bounty to fruition and they are poised to do more.

As farm member Martin Novom explains it, "our goal is not just to raise food, but also to raise consciousness. . . . We don't have all the answers, but we do have some of the questions. How are we going to continue to have sustainable farming? How do we save not just the soil, but also the farmer? Corporate agriculture is not the answer."

The Community Farm

The Temple-Wilton Community Farm is a model that other farms and communities would do well to study, for it may demonstrate a way to revive local agriculture and the plentiful social benefits that come from it.

*Reprinted from the original *Farms Of Tomorrow*

In January of 1986 Trauger Groh, who had recently moved to the region from a community farm in Germany, met with Lincoln Geiger and several other families who shared the dream of supporting local farms for local people. Dozens of social and economic factors stood in opposition to such an undertaking. Still, they felt they could make their dream real and so they decided to try.

Here's the basic idea they came up with: independent families in the area who wanted to would join together in an association known as the Community Farm. Some of the members had land suitable for farming, most did not. Landholders would make their fields available to those who were able and willing to use them. The rest of the people would continue their lives independently, but through their association in the Community Farm they would receive food from the land. In order to make this possible, they would support the farm by providing finances.

As Lincoln Geiger explains it, "many people don't want to use the land they have, but they would like to see it farmed. They make their land available so that farmers without land can care for it in their name. Under such an arrangement no one gets rich, but then again, no one goes hungry."

Of the approximately sixty-three families who belong to the farm, three individuals in particular work on the land with apprentices. The rest of the families live and work in the surrounding area as teachers, writers, builders, and so forth. They are part of the farm financially and in their hearts, but they all have private lives.

At the start of the growing season the farmers estimate their expenses in detail, including the costs of seed, salaries, tractor repair, diesel fuel, baling twine, and so forth. They present this budget to all the families in the community farm—and then collectively they meet the budget. Each family pledges how much it can pay in monthly installments. Families which have much give more than those who have less—not according to a formula, but according to each family's sense of what is affordable and appropriate. They work it out among themselves. Pledges have generally ranged from thirty-five to eighty dollars a month, though the farm membership has debated the advisability of establishing a minimum monthly pledge of fifty dollars, which would be $600 a year for a family. These pledges are totaled and applied to the farm's budget.

In 1987, for example, the budget for the Temple-Wilton Community Farm was $53,000. Together the families were able to pledge only $51,500 so the budget had to be cut back a bit. The money gave the farmers working capital and some income to meet living expenses.

The budget increased in subsequent years, but so did the number of members.

Under this model the farmers are not tenants but instead have an ongoing right to use the land. The Community Farm itself owns nothing. All land, equipment, buildings and animals are owned by the members individually. Members who own farm property are compensated only for their costs: taxes, depreciation, maintenance, and so forth. They make no profit, but they keep their land open and productive, the land is improved by the farmers, and the owners serve the community and their families by making it possible to grow fresh, clean food.

To create a land base for their community enterprise, as of 1990 three independently owned farms pool their resources: Echo Farm in Temple, The Temple Road Farm in West Wilton, and Plowshares Farm in Greenfield. Out of about 200 acres total of forest, pasture, and field, they produce nearly all the vegetables for the sixty-three families who are involved: carrots, beets, onions, parsnips, rutabagas, turnips, lettuce, spinach, chard, Oriental greens, kale, cabbages, broccoli, cauliflower, leeks, celery, tomatoes, peppers, potatoes, squashes, apples, cider, blueberries, eggs, flowers, herbs, milk, and meat. They hope to begin producing their own grain soon.

All the food that goes into the Community Farm store is available to the families. Rich or poor, they take what they need. If you are hard up for money, you still get food. The community knows you and helps you.

The agreement which binds the families together may be unusual by modern American standards, and it may still have some kinks in it, but it has worked. As farmer Trauger Groh sees it, "We need healthy farm organisms not just for the food, but also for our social well being, and for the education of mankind. Farming is so essential that one has to do it at any cost. We can stop making sewing machines or VCRs and life will go on, but we can't stop farming."

Aims and Guiding Principles

In February and March of the first year (1986), about twenty families met frequently to shape the Temple-Wilton Community Farm. Together they wrote a document setting out their intention, underlying concepts, three aims which would guide their actions, and also seven principles of cooperation.

The community farm, they wrote, was born out of their intention to unite their efforts and their land into one organism in order to serve the

local community with bio-dynamically grown food. They established
the following concepts:
- Landholders—The landholders give the members of the community
  farm, individually and in association, the right of the agricultural use
  of all their land, farm buildings, farm animals and farm machinery,
  except their homes and house gardens. This right is given free of
  charge, but with compensation made for all costs of the property:
  land taxes, insurance, depreciation, and repairs.
- Farmers—By taking over the right of use of the above-mentioned
  land, and with it the responsibility for the good agricultural use of
  this land, all members of the community become farmers. They
  either enact their rights to farm directly, by actually planning and
  doing the farm work, or by letting those members who have the time
  and skills do it, farm in their name. Those members who do the
  planning and the farm work on an ongoing basis and as a main occu-
  pation are called the "Active Farmers."

Spiritual Aims of the Active Farmers
- To make life on Earth possible ever again and every year anew in
  such a way that both the individual and humankind at large can live
  towards their spiritual destiny.
- To make land use and working of the land a way of self-education in
  the sense that a better understanding of nature can lead to a better
  understanding of man.
- To create the farm organism in a way that the above becomes possible and that it is made available in a therapeutic way to those who
  suffer from damages created by civilization and from other handicaps
  that need special care.

Legal Aims
- To make access to farm land available for as many people as possible.
- To create forms of cooperation that exclude employment and any
  form of paid labor.

Economic Aims
- To develop a natural organism so that it reproduces itself better and
  better and becomes more and more diversified, so that it can be a
  primary source of food for the local community.
- To achieve that reproduction and diversification with the help of the
  forces of nature inside the farm organism so that it becomes less and

less necessary to introduce into the organism substances and energy from outside and so that human labor is used as economically as possible.

- Individual profit through farming is not an economic aim of the farmers.

Principles of Cooperation

The farmers agreed on certain principles to make cooperation in the agricultural community possible:

- Everyone is individually fully responsible for his or her doings and their consequences. To enable others to help him in his initiatives and the consequences thereof, the individual must, in a timely way, let the others know what he intends to do.
- The individual incurs expenses to serve his initiatives. The expenditures made by the single individual increase the costs for all the others. Therefore, the individual, in cooperation with other individuals, has to declare what costs he projects to fulfill his initiatives within a certain time. The declarations of those that intend to spend money, combined together, make the annual budget for the farm. This budget has to be approved by the assembly of farmers (all members). Once the budget is approved, the single farmer is free to spend, throughout the year, the amount of money he has in the approved budget.
- Every farmer who spends money must be ready to keep books and records of such expenditures. The farmers agree on a scheme of categories in which the expenses are accounted for. The books have to verify annually how far the economic aims have been achieved.
- Every farmer (or member of the farm community) can leave the community at the end of the year when that member has paid his or her part of the annual cost.
- Every farmer gives all the other farmers of the group the right to substitute for him in his work if he fails to do or complete something he has taken on.
- It is understood that the better the cooperation among the farmers is working, the less goods and services are brought into the farm organism by individuals at the expense of all others. It is understood that the least desirable thing that should be purchased from outside is human labor.
- All farmers have to take care that they spend enough time on observation, planning and communications.

• The motivation to do things on the farm should come more and more out of the spiritual realm and less and less be directed by solely financial constraints.

Applying Common Sense

Anthony Graham, one of the farmers, says the biodynamic approach they use at the Temple-Wilton Community Farm is much misunderstood. In the main, the techniques are common sense and the application of common sense. To illustrate this point, he explains some of the practices they use to grow crops.

They begin the farming process in the depths of winter when the farmers think through their budget for the coming year, and then meet with the families to build a picture of what will grow through the summer. With discussion and active imagination they create the focal points that, in the growing season, they will bring their collective will to bear upon. In support of the farmers, the families pledge to meet all the expenses for the year.

When it comes time for planting, the farmers match the rhythm of their work to the rhythm of the seasons and the planets. For example, biodynamic farmers understand that the phase of the moon is revealing something about its relative ability to influence water on earth—not just the ocean tides, but also the pulse of the fluids in seeds and plants.

As for fertility, compost is the backbone of the operation. Biodynamic farmers don't just spread raw manure on their fields. Instead they compost the manure with straw, other organic matter, and herbal preparations. They use farm equipment to mix the compost so that the materials blend together in a harmonious way. They nurture the microscopic life in the material as it gently cooks over a season in preparation for its return to the soil, part of an unending cycle of birth, growth, and decay.

Biodynamic farmers are fond of spraying their crops, but not with chemicals; they use mineral and herbal preparations. For instance, they spray water mixed rhythmically with specially prepared ground silica, a natural crystalline material which has an affinity for light. Silica allows the plants to work more efficiently, to photosynthesize more effectively.

According to farmer Anthony Graham, the community farm has few problems with insects and disease. That's because they work hard to build good healthy soil, which gives rise to strong plants which can resist disease and pests.

Their next line of protection against insect damage is to spray herb teas. For example, in 1987 when the farm faced an infestation of striped cucumber beetles, which have an inordinate fondness for juicy squash leaves, the farmers sprayed a tea made of pennyroyal. It didn't kill anything, but the beetles found it disagreeable so they vacated the fields for a few days. That disrupted their feeding and breeding cycle. Since the spray wasn't toxic, it harmed none of the insects which prey on the striped cucumber beetle. Consequently, the tea brought the problem to a gentle end without disturbing the vibrant life in the fields. When pests do come, the farmers rely on hand picking, feeding the bugs to the chickens as a part of the farm's internal recycling.

As these examples illustrate, biodynamic farming is labor intensive. It requires much more work than farms which use chemicals to control weeds and pests. Trauger Groh observes that "biodynamic farming needs many hands, many people. You cannot employ laborers because they have gone into industry, and with present economics you couldn't afford to pay them either. Farming does not work well with employees, anyhow. It requires a far deeper level of interest and commitment."

Nowadays the Temple-Wilton Community Farm offers places for apprentices, people who are willing to take a step forward into something new, something that could help to preserve the earth and the farms. "You don't get paid by the hour," Anthony Graham says, "but then members don't pay by the carrot."

A Principal Challenge

Such an approach to farming calls for dramatic change. But then, if we are to keep our air and water pure, if we are to preserve the remaining farms, and if we are to create a culture which pulls communities together instead of fragmenting them, dramatic change is necessary.

One of the principal challenges of modern farming is, of course, economic. Lincoln Geiger says "if people were to pay the true cost of running a healthy farm, they would have to pay much more. Right now humanity as a whole is paying for it because we are expending resources. We don't pay to maintain the land or the internal economy of the farm. We buy everything from outside and bring those resources to the farm to produce. We need a more labor-intensive agriculture, not a more material-intensive agriculture."

The Temple-Wilton Community Farm found itself facing a serious economic challenge as they came into the winter and spring of 1989. That winter blew in cold and dry on the heels of scorching summer heat and drought in 1988 that reduced crop yields significantly. "It was a poor harvest," Anthony Graham commented, "we had barely enough to get through the winter."

Members of the community also faced the loss of a newly built school, which was destroyed by fire in October, 1988. With the school disaster and the poor harvest in the back of their minds, the membership of the farm came up about $13,000 short when they pledged to meet the farmers' 1989 budget of $68,000.

From the beginning, the farmers had emphasized that when you joined the farm you were not just buying vegetables, but rather pledging to support the farm—good years and bad. But when the reality of a small harvest hit home, some families withdrew. And so the farm faced a serious crisis.

In response to the crisis, members of the farm formed a task force and developed several innovative ways not only to cut the deficit, but also to reduce the work load on the farmers. "Now, for the first time," Anthony commented in 1990, " it has really begun to feel like a community farm. More and more people are plugging in their various skills to make this work. We feel okay about it. The crisis really galvanized people. We have lots of help now, and we are going to make it."

For a community farm to work, interest must extend beyond the land into the realm of human relationships. Because of the cooperation required, the members must be willing to know each other intimately and to work out the agreements and disagreements that naturally arise. Such contacts are the grist of community, and they are also what's missing in today's food system: by and large, our relationship with our food and the people who grow it is remote and anonymous.

That there are many people interested in change is evident in the waiting list of families who would like to become part of the Temple-Wilton Community Farm. But the farm is not ready to expand. A farm can serve only so many people. There are limits. For the concept of community farming to grow, other farms and other families must create their own organizations.

As Trauger Groh sees it, that is essential. He says the community farm has no future without a network in New England of 100 or so similar farms that can support each other through trade and association. Such a network, he says, has developed in Europe. "We want to develop a diversified program of foods that meet the needs of local communities and to produce it while the input of energy, materials and labor is reduced."

What the community farmers suggest has far-reaching implications. Agriculture has been the heart of the nation for two hundred years, but now the heartbeat is growing weak and irregular. Americans have left the land physically and perhaps spiritually as well. As a result we face grave problems. The community farmers are experimenting to see if they can't begin to show the way back and the way forward at the same time.

The Farm Revisited

Over the years hundreds of visitors have come to the Temple-Wilton Community Farm to see the fields, to talk with the farmers, and to deepen their understanding of what they are doing. The farm has been through many changes, some of them traumatic, but it has remained basically stable, and has continued to evolve according to the vision of the farmers and the community. Eleven years after starting, the aims and principles they set out for themselves at the beginning remain as the guiding force for the farm.

As the farm stands in 1997, Anthony Graham manages the vegetable production and the farm store; Lincoln Geiger manages the main herd of thirty head of cattle, and oversees the production of yogurt, and cheese; Trauger Groh manages some laying hens, a flock of sheep, and the milking Devon herd, whose cream he churns into butter. Trauger also travels widely to teach, nationally and internationally.

Rather than a hierarchy, together they have established a collaborative management. "It's an interesting effort," Anthony observes, "to see that three such diverse people would hold this together over so many years. We have our disagreements, of course, but there have been no major upsets. All three of us have always seen the value of what we are doing together as more important than what any one of us is doing, or wanting to have happen. We are peers. We can and do speak with each other freely. No one thinks of himself as the 'boss'."

When Anthony looks back over the history of the Temple-Wilton Community Farm he recognizes 1989 as a benchmark, an extraordinary year. "It really stands out for success," he says. "We've had eleven seasons by now, and 1989 was the year we really got into gear as a community farm and did well with production and finances. We knew then we were on the right track. We'd had a terrible drought the year before (1988), and both Lincoln and I were moving our houses that year, so the farm suffered terribly. It was a crisis year. We had about seventy-five member households and twenty-five of them left. But we were reinvigorated by the season of 1989, and went up to eighty-five member households. We also got our watering system in place, though we did not need to use it that year.

"What has happened through the 1990s is a period of steady growth. We went very quickly from thirty-five members our first year to fifty the second, then to sixty-five and on into the eighties. Since then we have grown in steady increments to 110 member households. The limiting factor for us has not been the desire to grow more food to feed more people, nor a lack of people wanting to join (we always have a waiting list), but rather the amount of land we have.

"Interestingly, we are growing no more acreage now than we were back in 1990. We are still at five acres, which is as much as our limited land resources will allow us, with some land for crop rotation. The amount of land that we have been able to cultivate in a given year has remained steady, but if you look at our figures you see that our production has generally gone up. The potential became obvious to me through visits to other farms to see what they were doing—especially a visit to the Smith's Caretaker Farm in Williamstown, Massachusetts (see *Example 6*). What they have done there is really wonderful.

"I began to see what you could do in a small area with much more intensive practices. So I started to put more compost into less land. I cut back any places in the fields that were not productive—places that were wet, or at the edges of the fields. I was wasting a lot of time before. I began to really focus on the most productive areas, and to bring those areas of the fields that weren't doing well up to a higher level. We tried to find out what was going on in our fields, where before we had just gone on luck: we would open a field and apply some lime. But then we really did careful analysis.

"Also, over the years we have been using the same fields and pouring in extensive amounts of biodynamic compost—again and again. And picking out the stones, which is a New England tradition. All that has really brought those fields into a completely different place. One begins to feel a different sort of tone in the field. Something different is living in the field after a certain amount of time. We've reached the point now where many of our fields have that kind of tone.

"On the other hand, we've also reached the place where our resources in terms of fields are feeling somewhat depleted. Because as much as we put in preparations and compost, we are working the fields hard and getting a lot of production out of them. If you can't rest the field for two or three years, after say three years of heavy production, then it begins to feel as if you are mining rather than farming—actually extracting the essences of the soil to a degree that you may not quite be replacing. So, we've actually reached a point in the life of this farm where it feels more than it ever has before, that we are too spread out.

"In any given year we are farming on five or six different fields spread pretty far and wide in two different towns (Temple and Wilton). Some of the fields are up to three miles away. That's tremendously wearing on people and machinery. In general, it makes the farm feel much too stretched out.

"There has been a real yearning to have things together—the cows and chickens are up at Four Corners Farm, about six miles away. Lincoln is forever driving back and forth to tend and milk the cows. All the compost is up there and has to be moved back here to be spread on the vegetable fields. It's an enormous strain on the whole organism of the farm—one even has to question what kind of farm organism this is, that is so stretched out. So this has driven us very strongly to the point of crisis. We must have a place where the cows and the vegetables are actually in the same place, even if we still have to use outlying hay fields. That would make an enormous difference in our lives, in the life of the farm, and in the budget of the farm."

As 1997 got underway the Temple-Wilton community farmers were looking actively from place to place, casting a net wide around the region to find the farm land and buildings that can serve as a permanent home.

Beyond this challenge, the farmers agree that there is enormous social strength in the farm. "We have 110 families, with a waiting list," Anthony explains. "People are knocking on the door all the time. The farm has proved itself. And there is enormous support from the community as well.

As Lincoln sees it, sharing the farm budget, sharing the food, putting out the newsletter, and inviting members, neighbors, and friends to the seasonal festivals have helped to strengthen the community. "Together," he says, "they create a tangible substance for the members. That's where the community meets the farm. And that's where they see what we are doing."

"I see the Temple-Wilton Community Farm as firmly established," Lincoln adds. "We know the details well. We know what's missing, and what we are working toward. In our case, the membership is a fairly professional group, and so they have little time to spare for helping, either in the fields or on the various committees. It's very hard to find volunteers. It was easier in the beginning. The group has changed over time. They are not so able to share their time with the farm. We connect as a community through the store, the newsletter, and the festivals that we hold through the year."

Financially, the farm still maintains an open system for member families. After learning the budgetary needs of the farm, people pledge what they can afford in support, rather than paying a set fee. "The amount that we need per family has remained fairly constant," Anthony observes. "As we have grown through the years the budget has grown, and so has the number of families, but there's been a fairly constant ratio of families to budget size. What we have needed generally over the years is around a thousand dollars per family, which works out to around eighty-five dollars a month. Most people pay by the month, though some prefer to pay a double pledge from June through November.

"The farm store is open all the time, and is still stocked twice a week round the year with products from the farm. We grow an enormous quantity of food that goes into our root cellar—carrots, beets, parsnips, turnips, rutabagas, celeriac, potatoes and cabbages—and we also have onions. Anything that is storable long-term, we grow and store, and then set it out in the farm store through the winter, along with milk, yogurt, butter, and cream cheese from our dairy, and bread, which is baked by one of the farm families (the Plowshare Bakery). Those goods come in twice a week to the store, and people can come and fetch it whenever they want.

"We've also started a little coop about a year ago, in collaboration with Northeast Coops. So we stock the store with things from there, and mark the prices up a little bit to help provide some income for the farm (a couple of hundred dollars a month). We stock items such as bulk grains, paper products, jams, dried fruits, and so forth. Even with our mark up, the prices are way below what you would pay at a health food store. It's just another service that makes it worthwhile for someone to be part of the farm and to keep paying their pledge. So, the average pledge of support is about ninety dollars per share per month, but there's still a tremendous range: some paying forty dollars a month, and some as much as $120 because they can, and because they want to make the farm possible for other people.

"I have continued each year to monitor our production in terms of its financial value, against the prices you would pay for similar produce at Northeast Coop, or local natural foods stores, or even the large, mainstream supermarkets around here. I take widely ranging price readings, and we always come out underneath all of them—whether its the wholesale price at Northeast Coop, or the local supermarkets, where you get vegetables grown everywhere, vegetables that are not even organic. You also have to remember that our members get a double premium, because we are not only organic, we are biodynamic—and the stuff is so fresh and so full of life force. You can tell the difference.

"In addition, people realize that they have the satisfaction of supporting a local farm, and making its existence possible. That was something that we used to talk about in the beginning, but people really get it now. They know they are doing it, and they know how important it is. The awareness in the country as a whole of what is going on in terms of food, and the need for organic food, is growing in leaps and bounds. People realize that they are lucky to have this kind of a situation on their doorstep."

Toward the end of the 1994 growing season there were changes in direction and philosophy at the Lukas Foundation, which had been a large and important part of the Temple-Wilton Community Farm from the beginning. Some of the newer co-workers at Lukas decided they did not wish to go on with the arrangement, and so the farm was abruptly forced to find a new situation, and to abandon fields that had been biodynamically worked and composted for fifteen years. The farm also lost its relationship with the handicapped adults who live at the Lukas Foundation, and who had worked with the farmers to grow the crops. The severing of those personal relationships was especially hard.

At the time, Anthony recalls, it seemed as if the end was near, especially as the 1995 growing season got underway. It proved to be the hottest summer on record. But the farmers and the community rallied. They moved the greenhouse to the Groh land, which was already part of the farm, and also built a new farm store and root cellar upon the land. The cows were moved six miles away to the Four Corners Farm, where they reside today.

"For us, the next step is to develop a situation where the farm comes together on the land, to consolidate physically," Anthony says. "This is the next big goal. Beyond that, anything is possible.

"As far as the larger context of CSA goes, I would say that we are reaching more and more people. Over time folks move on from this community to live elsewhere, and they carry the message and sometimes the initiative and the will to start other CSAs. The whole thing is poised at a point where it can really grow, especially as people become alarmed at the growing mergers and alliances between seed and chemical producers. As more people become aware, they will turn toward CSA for sustenance.

"We know now that CSA is not a flash in the pan, or some fly-by-night experiment. There is a demand for this. The only limit is the will of people to make it happen in various communities. I see enormous potential for the growth of CSA to continue."

From his vantage as an active CSA farmer, and also a board member of the Biodynamic Farming and Gardening Association, Lincoln also sees great potential. "I realize that the CSA movement is growing very quickly," he observes. "It's an idea of a reality that is very attainable today. Right now as far as the potential of CSA goes, I think the sky's the limit. There are many people wanting food, and there are many people wanting to grow food. CSA is a great medium to do that, for people to meet, both from the farmer's and the consumer's side. We have lost our way with nature, being a very urbanized culture. CSA makes a clear linkage back to nature for many people. You have some connection, some roots set down where the food comes from. It means a lot."

Lincoln Geiger–The Importance of Farm Animals to CSA

As with his friend and colleague Trauger Groh, dairyman Lincoln Geiger feels strongly that other CSAs should give serious consideration to having animals on their farm. He has considered deeply the costs and the benefits.

"First of all," Lincoln observes, "the budget goes up substantially when you get into animal husbandry. It's a large increase in costs if you are going to produce the fodder for the animals, because you need land and you have to invest in all the equipment to cultivate the soil, to plant, and to harvest. You need large buildings to store the forage, and a fairly substantial building to shelter the animals, depending on what kind of climate you are in.

"The other consideration is that many of the people interested in the CSA movement are vegetarians, and some are opposed, in various degrees, to keeping animals. . . . So, if you have already started a CSA with vegetable production, and you are trying to get a sense of what the membership feels about having animals, you will be prone to not getting a positive response. There may be some individuals who will oppose it.

"At the Temple-Wilton Farm we basically have maintained a line where we don't raise animals for the meat. We raise them because they belong on the farm to make it possible to grow vegetables, and to have milk products. At a certain point, though, the animal will have be butchered or have to die. You just can't fit that many animals on a farm. If you produce milk, then you need to have them produce offspring annually. The herd grows very quickly to the point where you have to either sell or butcher the animals.

"In my view we provide a place for cows to live, giving them the opportunity to evolve by living in a good way, eating natural fodder and ranging free on the pasture where they find shrubs, herbs, fragrant grasses, and clovers. They often calve in the pasture. If farmers give up cows they will have nowhere to live and will possibly become extinct. It is far better to share the land with the animals than to eliminate them. Of course, I condemn the misuse and exploitation of animals in factory farms. However, the fact that some people misuse animals should not lead us to bring about the extinction of those animals.

"So, those are some of the obstacles, the financial situation, and the resistance against having animals. Also, the land base has to be much larger. You need at least two acres per cow, if you're dealing with cows.

"But on the other side, the positive side, there's a lot to win, a lot to gain. First of all, you don't have to buy and bring in your fertility. If the membership is interested in dairy products, you can get a whole range of food: fresh milk, and cheese, and yogurt and butter. In my view, fresh, whole milk is an essential health source.

"Whole foods are becoming hard to find. More and more often our food choices are becoming fragmented and altered: low-fat, no-fat, empty starch, and isolated nutrients. The mentality which guides these processes and leads to these fragmented foods is undermining our health. The main problem with not having whole foods is the development of degenerative diseases. We have created a totally disunited world, both around us and within ourselves.

"The whole atmosphere that the animals create on the farm, especially for the children, is essential. We try to have a variety of animals, so it becomes like an ecological landscape. We have some sheep, and goats, pigs and chickens, geese, ducks and the cows, calves, and bulls. With all the different areas that they graze and feed on, and all the different places that we harvest forage for them, it creates a landscape for our local area, which is an essential ingredient of the image, or sense, the local people have of themselves and where they live. The members of a CSA that has a landscape such as this feel that they are part of and share in that landscape, too.

"Children have a deep experience of the landscape, and more intimately the farm. If they are members of a farm I think they will have something valuable to cherish as they grow older and think back on when they were little, and they were part of something like this, something important. This is an educational component of CSA for humanity. You hear adults talk about such childhood memories all the time.

There is a common and good feeling among people who visited farms when they were young. The feeling of belonging to nature, and having an affiliation with nature will be stronger if the children have these experiences—and adults, too. One sees that the land belongs with the people, together with all the different life that forms nature everywhere around us. The wonderful image of Noah saving the animals is again becoming a reality. Much is at stake and we are the keepers of the Earth."

Example 2

The CSA Garden at Great Barrington

The Farm 1990*

In the Berkshire Mountains of western Massachusetts, farming has become virtually extinct. Aerial photos from earlier in the century show the region was almost completely pastoral, but now, for example, there is only one struggling dairy farm left in South Egremont. Down the road in Great Barrington, the First Agricultural Bank hasn't loaned a cent for farming in over seventy-five years.

However, without any large pool of capital, a group of about one hundred families is demonstrating one way that farming may come back in the region. Together they have formed the CSA Garden. The acronym CSA stands for Community Supported Agriculture, an idea which has enabled the families to directly support some farmers economically, while the farmers produce an abundance of fresh, local food for the families.

Oddly, their story begins not in Great Barrington but in Switzerland. While living there in the early 1980s, a young American named Jan Vander Tuin learned of a new kind of food production cooperative operating in Geneva, Basel, and Vaduz. After studying them, he helped to establish a similar venture, the Co-operative Topinambur in Zurich.

The concept of these new cooperatives is simple: divide the costs of the farm or garden among shareholders before the growing season begins. Instead of an agriculture that is supported by government subsidies, private profits, or martyrs to the cause, they create an organizational form that provides direct support for farmers from the people who eat their food.

Vander Tuin brought the CSA concept from Switzerland to America. As he shared his passion for the idea around Great Barrington, a core group of people began to form, including John Root, Jr., Robyn Van En, and Charlotte Zanecchia. After the core group got together they contacted gardener Hugh Ratcliffe, and hired him to take responsibility for growing food on a large scale. Hugh had been a research biologist at

Cornell University, but he quit when he realized that he disagreed with the predominant scientific view of plant life as basically mechanistic. Instead, he began to study biodynamics. "We can't see the forces at work in the garden," he has noted, "but we can see the effects of working consciously with them."

Biodynamics and CSAs are well suited to each other. While agricultural methods which treat the earth as a living being have the potential to connect people rightly with nature, the community dimension of the CSA garden can likewise connect people rightly with each other. Instead of being confronted with all the usual compromises that a farmer has to make—all the business and marketing issues that have nothing to do with cultivating the earth—under this approach, the support of the community frees the farmers to focus on what they do best: farm.

The First Project

The first project the CSA Garden undertook was in an old apple orchard during the autumn of 1985. They began by selling thirty shares in the orchard, the cost of shares based on the expense of picking, sorting, storing and distributing 360 bushels of apples. They also pressed apples for cider, hard cider, and vinegar. These products were distributed among the thirty shareholders. In the minds of the core group, the success of this harvest established that the CSA concept was really going to work.

While the apple project was underway, the core group leased land for a large garden from Robyn Van En at Indian Line Farm in South Egremont, Massachusetts. The lease was for three years with an option to buy, and it included water, electricity, vehicle access, and the use of an out-building.

The group borrowed a tractor, and purchased a disc harrow, a wagon, assorted hand tools, and seeds. They also made arrangements to use a greenhouse for starting seedlings, and a cellar for winter storage space. In the fall of 1985 the land was plowed, composted, manured, harrowed, and seeded with a cover crop.

Over the winter the core group met and estimated that the average person consumes 160-200 pounds of vegetables a year, while the average house size is two to three people. Therefore, they figured one share needed to be about 400-600 pounds of vegetables per year, or an average of ten pounds per week. So production was planned according to these figures, while also considering averages of consumption for forty to fifty different kinds of vegetables, herbs, flowers and mushrooms.

Based on their estimates, the core group prepared and issued a proposal for community supported agriculture in the southern Berkshires, and asked for people who supported the idea to send them ten dollars. Only two people did. Despite this disappointment, they felt they could succeed, and so they went ahead with their plans. Eventually they sold about fifty shares to individuals and families that first year, enough to go forward. Each year since then, though, the CSA Garden has grown: 135 shares by 1988, and about 150 shares in 1989—which is about the garden's upper limit.

The cost of a share has changed from year to year. John Root, Jr. says "when we made our first budget we were worried about too high a price; a share was $557, based on paying every worker $7 per hour. So, in wanting to get the thing to work we priced the wage too low. We have been steadily working it up. A gardener needs to make $40,000 a year to live in the Berkshires. We are only talking about seven months of work, though, so that gets it down in the area of $25,000. In 1988, the gardener made $16,000. In 1989 he will make $20,000."

While wages have gone up, share size and cost have been moderated to meet the needs of the shareholders. In the 1988 season, for example, a share cost about $300 and entitled a shareholder to one large bag of fresh vegetables and fruits each week for the eight months of the growing season, and one large bag a month in the winter—potatoes, beets, carrots, and so forth. Over the long haul of the season, a share yielded enough vegetables for two people who are vegetarians, or for a family of four with a more widely based diet.

Nuts and Bolts

Up until 1989, the CSA Garden was an unincorporated association managed by its co-workers on behalf of the shareholders. The main administrative vehicle was a weekly meeting of the core group that shareholders are welcome to attend. Rather than a hierarchy with one person on top giving orders, the group has striven for consensus in decision making.

"Once we had the core group," CSA member Andrew Lorand has commented, "we were on the way. The important thing for anyone starting out like this is to get a core group of four or five people who are committed to making it happen. Later, when we sent out the prospectus for the garden in the spring, it was actually quite straightforward and easy."

Andrew says community supported agriculture (CSA) is a simple concept. Because of economics and politics it usually isn't possible to do the right thing in the garden. But he says the CSA Garden takes as its starting point, doing the right thing in the garden. "We have a responsibility towards the way that we treat the earth, and only by treating the earth in as comprehensively good a way as we know how, can we expect it to properly nourish us. If you take that as a starting point, then that's something people can connect with. It's a practical ideal. For that to be possible, though, there has to be a community that says 'we want it.' Because if you do it the other way around—'I'm going to have an organic garden because I can get a higher price for my produce'—forget it. It will never happen. There's nobody to take the responsibility. So CSA basically says, if you want to take control of your own food situation, then join us and support the gardeners. Then the gardeners will support you with clean food."

Each year over the winter, the core group draws up the budget and calculates the price of a share by dividing the costs of producing the harvest by the number of shares the garden can reasonably be expected to provide. In practice, the budget, which reflects the experience of the previous year, is the basis for the estimated price of a share. However, if they underestimate, they can charge up to ten percent extra. Should there be a financial surplus, which the group concedes is unlikely, it would be refunded to the shareholders.

In the spring, the core group issues a prospectus for the coming season, showing the shareholders what the costs will be, and what they can expect in return, including a chart showing approximately when particular crops will be harvested and made available. The prospectus states, "our gardeners plant to provide a balanced harvest each week. Since there are so many variables in nature, surpluses and deficits of particular crops are likely."

The farm harvests from May until mid-October. For the first three years, subscribers had the option of picking up their shares either at the garden, or at the Berkshire Coop Market in Great Barrington. In 1989, though, the CSA decided to experiment with home delivery. They reasoned that since people were just coming to the farm, picking up their vegetables and then leaving, the producer's group might just as well send a truck around to deliver the food—it would save energy and cause less pollution in the long run.

The development of the CSA Garden over the first three years has been solid. The vegetables have been of high quality, the amounts have been, for the most part, more than adequate, and the distribution has worked satisfactorily.

In the beginning, the CSA Garden intended that each shareholder would do two days of work each year in the garden, but that just didn't work out. People were supportive of the CSA concept, but they didn't get directly involved. The group found it was up against many strong cultural currents—the lure of the electronic cocoon. This reality prompted much long discussion within the core group around the question of what they were actually doing: creating a community of consumers, or simply marketing a somewhat unconventional product.

By creating their own market, the CSA Garden has escaped the bind of the vegetable market in which farmers must often sell their products for less than it costs them to produce it.

An Open Secret

One noteworthy facet of the CSA Garden is the involvement of mentally handicapped people from the nearby Berkshire Village. They participate not just as consumers, but also as workers in the garden.

John Root, Jr., who was active in forming the CSA, has also been a leading force in the involvement of handicapped people with the CSA. He says his basic interest is in community, and working with handicapped people. "When you are working with handicapped people you are creating, whether you want to or not, a community around them, a community in which they can live. The handicapped person tells you something about what it is to be human, because we usually think in terms of 'I'm intelligent, or I have particular talents.' But the handicapped person doesn't have that. In creating a community in which a mentally handicapped person can live a decent dignified life, you also discover that that's the way everyone should live. This is an open secret in which everybody is gradually discovered.

"My particular interest has been in creating situations where everyone in the community is responding to the needs and the possibilities, and where people are not being coerced. As I see it, if someone hires you to do a job the salary is a coercion. You agree to do the job for a salary; now maintaining the job by doing things you don't necessarily think are right or believe in are all related to the fact that you are getting paid a salary. One of the things, for instance, that a mentally handicapped person can't do is to support himself with a salary, a wage. This raises all kinds of questions."

Principles and Pressures

The members of the Great Barrington CSA started out with a long list of ideals, most of which they have held to fiercely despite ups and downs, the inevitable conflicts among people, and the constraints of the modern world. Among those ideals are respect for the workers as well as for the earth, plants, animals, environmental limits, and cycles of nature. The CSA has also made a commitment to use organic and bio-dynamic methods, to be energy conscious, to maintain decent working conditions, to emphasize the therapeutic value of agricultural work, and to support community control of land.

What has been a problem, as even their own newsletter states, are philosophical differences among the members of the core group. The problems have arisen in three areas: the relative importance of biody-namic gardening techniques versus standard organic techniques; deter-mining who has how much decision-making authority; and determining the role of the shareholders in the CSA. Such issues are likely to arise in any group enterprise.

The CSA Garden has also faced the challenge of land economics. According to John Root, Jr., "the single biggest issue in the development of CSAs is acquiring land. So far all CSAs either have a donor who has given them the land, or someone who is very sympathetic who is practi-cally giving them the land, or they have a very tenuous agreement about the land. . . . In the long term, we need to create a consciousness that land for agriculture needs to be taken out of the normal economic equa-tion. We need a CSA land trust—angels who will buy land for the CSA. But one way or the other, the agricultural land must be taken out of the marketplace." (See *Appendix B.*)

For the CSA Garden, an hour of reckoning with the question of land came at the end of the 1988 growing season. Their three-year lease with the Indian Line Farm expired and the core group faced the questions of whether to buy the farm, or move elsewhere, and whether to form a cor-poration of some kind to do this. Since they make decisions based on consensus, this was not an easy process.

"We knew in the very beginning that we needed to have a reasonably secure garden site. We felt reasonably happy about Indian Line Farm," John Root Jr. says, "because we had a three-year lease with an option to buy. But we did not succeed in buying the farm because we couldn't agree on price."

Ultimately, after much hard work and discussion around these issues, the CSA Garden arrived at plan for the long-term future. Specifically, the members committed themselves to establishing a comprehensive farm that would support its community for years to come, an agricultural enterprise farm that would provide quality food for consumers, and a healthy livelihood for farmers.

The CSA Garden group intends to form a land trust which will acquire suitable land either through a ninety-nine year lease, or ownership under a land trust arrangement. The group considers that land prices are inflated due to speculation, and they want to take some suitable agricultural land out of the economic rat race.

The group envisions a full biodynamic farm of 100 to 200 acres serving the complete nutritional needs of 200 to 225 families with milk, butter, yogurt, cheese, eggs, chicken, lamb, and beef, as well as fruits and vegetables.

As Andrew Lorand puts it, "we're convinced that in the long run the care of the land as well as the ability to meet the total nutritional needs of the consumer demands a whole farm. And the ideal, of course, of biodynamics, is to have a whole organism which takes care of its own needs. . . .The other thing that we are striving towards, which is perhaps a bit different than some of the other CSAs, is being able to meet the total nutritional needs of the members of the community."

A Plan for the Future

To help realize their larger vision, in early 1989 the CSA Garden reorganized internally, forming two distinct groups: the producers and the consumers. In this way, they hope to activate and maintain a supportive dialogue between the two. "You need two clearly defined groups to have a dialogue," Andrew comments. "Otherwise you have a monologue among many."

The producers will carry the farming initiative while the consumers will take some of the risk of farming from the producers. The producers association will own the means of production, and also the animals, buildings, machines, seeds, and so forth. And then, as they settle onto whatever land they acquire, the plan calls for the producers to also own living quarters, gaining equity over time.

Their consumer's group has elected to incorporate as a non-profit organization, while the producer's group will remain an unincorporated association for the time being.

To keep the garden rolling through the year of transition, the CSA Garden extended its lease with the Indian Line Farm through 1989. Meanwhile, the members of the CSA Garden have begun working with Professor Christopher Nye at Berkshire Community College. He has developed a Farm Match Program, to match young farmers in the area with people who have land but who do not want to farm it. Together they intend to locate a piece of land where they can develop the comprehensive farm of their dreams.

The Farm Revisited

The CSA Garden at Great Barrington has been through many changes since 1990, including a name change to the Mahaiwe Harvest CSA. (*Mahaiwe* is a local Native American word meaning "meeting place".)

According to founding member John Root, Jr., along about late 1990 the community recognized even more strongly the need to secure land for the farm on a permanent, or at least long-term basis. When the group was unable to reach a suitable long-term agreement with Indian Line Farm, its first home, it began a search throughout the area. Eventually, they found a fifty-five acre parcel, with a house and barn, in a Great Barrington neighborhood known as Housatonic. The owner, Mrs. Levant, wanted to keep the land open for agricultural use, and was willing to sell it for its agricultural value, estimated at $300,000, rather than the market value, which was much higher.

The CSA formed the Mahaiwe Harvest Land Trust as a legal vehicle to enable the transaction and long-term care of the land. The group then engaged in a lengthy process of planning, negotiation, and fundraising, working closely with the American Farmland Trust (see *Appendix F—Resources*). With the assistance of the AFT, they were able to secure for Mrs. Levant a $167,000 agricultural easement from the Commonwealth of Massachusetts. The terms of this easement ensured that the land would always be kept open for agricultural purposes, and the money dramatically reduced the amount the CSA needed to raise for the purchase. Ultimately, after much work, the group came up with a down payment, and then converted the house on the land into three apartments for the farmers. The apartment rent the farmers pay is used to make the monthly payments on the rest of the mortgage held by the Mahiwie Harvest Land Trust.

At this point, in 1993, Hugh Ratcliffe moved on to farm elsewhere, and David Inglis took over as the farmer. As of 1996, the Mahaiwie Harvest CSA had a thriving membership of 150 households, and an associated "market garden" which grows greens for local restaurants. The CSA members come to the farm twice a week to pick up their shares. "It's changed a lot, but it's still going strong," John Root, Jr. observed. "We no longer have regular festivals at the farm the way we used to, and the farmer has taken over many of the responsibilities that the Core Group used to handle. But we are still a CSA, and we are still supporting a farm, and producing a lot of good food for the members."

According to farmer David Inglis, he and two apprentices are actively cultivating about ten acres, keeping about half of that in green manures each year. They plan to add four more acres to their rotation in 1997. The farm produces a full range of vegetables, and has a popular U-Pick program for blackberries, raspberries, strawberries, and flowers. They are negotiating with other CSAs to bring eggs, milk, and other dairy products to the CSA membership.

The Mahaiwe Harvest CSA also has a buying club, that works much like a typical coop. Through the buying club, CSA members are able to purchase bulk grains, herbs, supplements, cosmetics, meat, bread and other organic products at a typical discount of thirty percent off the prices in a retail store.

Example 3

The Brookfield Farm

The Farm 1990*

It happened quickly. The farm across the road from David and Claire Fortier began splintering into two-acre house lots. Seemingly overnight, suburbia took root in fields which had supported crops for decades.

As they watched the houses go up, the Fortiers became deeply concerned about the loss of the farm land in their neighborhood of South Amherst, Massachusetts. Thinking back on that time, David, a retired professor, recalls "there were about seventy acres of land here which had been abandoned by the original farm around the turn of the century. The land had been apple orchard, hay fields and pasture. But it had been allowed to overgrow for about sixty years, and then for the last twenty it was rented out for growing corn, and they mined the soil. They'd just throw in some fertilizer and some corn seed, and then take the corn out. They did that year after year, and there was not much left in the soil. When we arrived here about 1963, it was still in corn. They left the field fallow for a year, hoping that would do some good. But nothing would grow in it, not even weeds. It was like a desert with sand and the topsoil all blown and washed away."

The property eventually came up for sale, but at a premium price. So it looked like the land was going to be developed. Ultimately, though, David and Claire were able to purchase the land. But what then? They were not farmers. To keep the arable acres clear, they continued to rent it to a farmer who grew corn. But they were looking for a permanent farmer who would treat the land with full respect. For a long time, without luck, the Fortiers advertised for a farmer.

While this was going on in South Amherst, Ian and Nicolette Robb were living on the other side of the Atlantic in Aberdeen, Scotland at the Camphill-Rudolf Steiner School for handicapped children. Ian was apprenticing in the garden, and Nicolette was completing her studies in curative education.

130   *Reprinted from the original *Farms Of Tomorrow*

Eventually Ian enrolled at Emerson College in Sussex, England, in the biodynamic agriculture course. Then in the spring of 1981, the Fortier's son, Thomas, visited the college and pinned his parent's advertisement for a farmer on the bulletin board. Ian and Nicolette saw the ad. A short, two-month correspondence followed, and by August the Robbs had arrived to begin farming.

Hatching a Plan

"When we arrived," Ian explains, "all that was here was the Fortiers and their hopes of having someone work the land. So it was just the four of us. We felt our first task was the land, to really deal with the land and work out a plan of what we should do. We took soil tests all over and it was a horror story because of the continuous planting of corn. The land was in poor condition, hardly any organic matter, terribly low pH, and so on and so forth. So we could see that before we could grow anything here we really had to improve the land. I mean, there were no animals, nothing, just land. A small barn had been built, and there was an old tractor and a few pieces of equipment. We had a unique opportunity of pretty much starting from scratch, and we could do what we felt needed to be done."

The Robbs looked around and saw the obvious. Amherst is located at the easternmost edge of the Connecticut River Valley in Massachusetts, one of the top vegetable growing regions in the country with highly fertile soil. There are dozens of vegetable growers, and the Robbs realized there was no way they could even come close to competing with the cosmetic quality or price of their produce. Vegetable stands seem to sit at every bend in the road.

"We felt that for something to come of our endeavors, it had to be based on a fertile, healthy farm." Ian explains. "We were facing some serious economic questions. We weren't a trust or a foundation then, we were just a group of four people. David and Claire had pledged to stand behind the farm financially for a few years to get it going but we still had to face the questions. What shall we do?"

The Robbs and the Fortiers had many meetings, just the four of them, to decide how to proceed. Right from the beginning they wanted to build toward a complete biodynamic farm. They had the whole-farm concept: the cows, the chickens, the grain, the crops, and so forth. But they had to start from virtually nothing.

"We had been gardening biodynamically for seven years at that point, so we had some experience and lots of theory," Ian says. "We knew where our particular talents were: berries, raspberries and strawberries. It looked like we could plant those berries here. We had grown vegetables all of our working life, but with so many vegetables available in the valley we had to wonder. Was there really a place for our vegetables here? We were a little bit cautious, let's say, about suddenly making another vegetable garden here. The organic market was virtually nonexistent, so we weren't kidding ourselves that suddenly people were going to start lining up for our produce."

Nicolette explains that they had to learn about the microclimate of the farm, which is in the shadow of Long Mountain, a frost pocket, which shortens their growing season. "We always started small," she says, "almost experimentally, and then went from there. As I look back on it, that was probably a wise move; we never made any huge mistakes as a result."

The Robbs started in 1982 by creating a huge garden to grow food for their own family. The garden did so well that they had a super surplus. That's just when the Bread and Circus natural foods supermarket opened, a Godsend from the Robb's perspective. The store began to change people's awareness of the importance of natural foods, which suddenly became important to people. Interest blossomed. And as it did, Bread and Circus began to purchase the surplus crops from Brookfield farm, as much as they could get. Soon their market garden was off and running, successful and developing a reputation for high-quality fruits and vegetables. But the Robbs were not satisfied.

From Market to Community

As they settled in and developed a feel for the community, it became clear to the Robbs that many local people had no access to the farm. Nicolette says "it was, for us, a feeling that this was just not right. At this time the farm was still the Fortier Farm, and we were still a market garden selling our produce to Bread and Circus or to a distributor in New York City. It just did not sit well for either of us. We had an underlying sense that something was just fundamentally wrong. And that's what induced us to talk with Trauger Groh of the Temple-Wilton farm, about sixty miles away in New Hampshire."

"With Trauger we shared our frustrations, our sense of what we wanted. We were just not able to practically put it into effect. He was the one who was able to help us to bring it down into the nuts and bolts of a working operation. We needed the help and the knowledge and the experience that he already had. When I look back at all the diverse elements coming together as they did, it just seems so right."

With Trauger's help, they decided that the way to create the full farm they envisioned was to link it directly with a community of people who would pledge annually to support the farm. This was the genesis of Brookfield Farm. The concept is simple but in some ways revolutionary.

Beginning in June, 1987, fifty-one households joined together to support the health of the land, its crops, the overall farm, and the newly formed community. They came together out of a shared understanding that agricultural problems arise from an imbalance between the agronomic, economic and social realms. They believe the earth is ailing, and that, together, they have the capacity to take responsibility for a portion of it.

One of the foundations of Brookfield Farm is the realization that, in today's world, not everyone can be on the land. The independent families who make up the farm community entrust the care of the land and crops to people skilled in the art of farming. The farmers' financial needs, as well as all other farming expenses, are covered by the monetary contributions of the farm community membership.

The families who have chosen to associate with each other as Brookfield Farm pledge support each year, providing the financial underpinning necessary for the farm's operation. In return, the farmers and farm provide, without cost, not only the vegetable and berry needs of the community, but also the possibility for community members to have a direct connection with the land, the farmers, and the food they eat.

"We were in with faith the first year," Ian says. "We formed a small core group and established a scenario. We said, well, let's start the community farm. We don't know how many members we'll get, so we'll keep the wholesale business going, so that if it doesn't work out we'll be covered. We'll go on selling to Bread and Circus and sell at a farmer's market. But we had a strong feeling that maybe we should burn our bridges. That the community farm might never take off because we didn't have much faith in it. You know, if you really had a tremendous amount of faith in it, maybe that would help. It was scary. But what we did was to say that we would go 100 percent to the community farm and that we would have faith that there would be the families we need. Of course, we had been talking about it and we already had interest from some families, but not the full complement we were hoping for. So we went ahead on faith."

Nicolette adds "it didn't take long, really, for us to realize that you cannot do justice to both operations—a community farm and a market garden—at the same time. We knew that ultimately the market forces would win out. To succeed in the market requires extremely aggressive selling. You have to keep your focus on the market, and you could not give your attention to the community and that dimension of the operation. There's just not room for a dual enterprise."

How It Works

Each year in March pledge forms go out to the approximately fifty families who are part of Brookfield Farm. The pledge amounts are based on the estimated budget of all farm expenses for the coming year. As both Ian and Nicolette aver, the pledge does really not buy a subscription to the harvest, but rather commits a certain amount to support of the farm. In return for that the families do get produce, but they have pledged to support the whole farm organism, not just the vegetable garden. "When you choose to support the farm, it's different," Nicolette says. "If you are just paying for some new way to get vegetables, then what happens when the farm has a bad year? Do you withdraw your support? That's when the farm needs it most."

Ian tells how in 1988 they had a deficit between the cost of running the farm and the pledges they had taken in. "Community members got together," he said, "and started brainstorming, 'what can we do?' And different activities happened, fund raisers. It wasn't 'Ian and Nicolette have a problem,' it was 'our community farm has a problem.' There were tag sales, and one community member had a restaurant in town, and if a community member went there for a meal then ten percent of the bill went to support the farm. People saw there was a need and actively responded. We can't generate money here. We can't say well, 'we'll sell off an acre of carrots to make some extra money.' We're feeding people, and we can't make money. There's nothing to sell." Reflecting on the way the community members responded to the crisis of 1988, Ian says "it was heart-warming. It made us feel 'this is what it's all about.' We recognized that there are people out there who are as concerned about the farm as we are, as the farmers are."

People come to the farm for their produce, where it is arrayed on tables. Community members take what they feel is necessary for the week, but the member in charge of the shop assists them with information on what crops are coming in, or which crops are in short supply— information about what is going on in the garden.

The people who make up the farm community give financial support to enable the farm to function in a radically new way, and they also have an opportunity, if they so choose, to support the well-being of the farm by helping out, from time to time, with chores like weeding or thinning crops. The farmers devote themselves to maintaining nature's balance, supplementing the soil's fertility, producing the highest quality crops, and building a healthy farm without being dependent on production for profit. The community members not only partake of as much or as little of the produce as they need, but they can also appreciate the intimate union between people and earth from which, in society's present structure, people are often divorced.

Through financial pledges the community commits itself to paying whatever is necessary to cover the costs of running the farm for the year. With this commitment, the possibility exists for an increased variety of produce and the ability to provide for more and more of the members' needs each year.

As Ian puts it, Brookfield Farm has "taken economics out of the center. You know: 'give me the carrots and here is the money.' That is all taken care of by the treasurer and it happens in advance so the farm is taken care of for the year. Already we have separated the idea that the produce that you get at the shop is worth some equivalent of what you would pay at the store, or something like that. No, it's because the whole farm was made possible, right down to the pig in the forest, or the cow in the shed. Because all of that is possible, there is produce on the table. So we keep on emphasizing that how you pledge on an annual basis, the guideline for that is derived by the total budget somewhat divided—and it may not be equally divided—by the number of adults in the community. And that's not an unlimited number. We can only feed so many. So the farm is really almost tailor made to the community that is supporting it."

Farming in the Forest

Brookfield Farm is situated above the Lawrence Swamp aquifer, the major water source for the Town of Amherst. Out of seventy acres total, approximately fifty-eight acres is woodland in the swamp, so it's a very sensitive area. Without chemicals of any kind—for which the town is deeply grateful—the Robbs produce an abundance of fruits, vegetables, herbs and honey.

Their hay fields, pastures, and forest land are arranged as if in concentric rings. The heart of the farm is the four acre vegetable garden. That is surrounded by the hay land, which is a little less intense in terms of activity, and then the forest land. The Robbs are steadily moving the farm into the forest. Ian explains: "We are seeking to permeate it completely for the health of the forest, which has suffered from acid rain. Livestock will be allowed to roam in the forest, pigs and cows. They graze on some plants and trees, and trample others. They also fertilize the soil with their manure. And eventually, over the years, the forest land begins to take on a park-like quality. We feel that this is where you can have a mutually supportive relationship with the forest. No longer just using it as a supply of wood, but the farm can, through a diversity of livestock in the forest land, start to give something back."

It was with forest grazing in mind that the Robbs chose a particular breed of cow, Irish Dexters, which graze and will do well in this setting. The Dexters can be used for beef or they can become milk cows. The Robbs intend to use their cows for beef now, because they are not yet in a position to set up a dairy.

Keeping cows in the forest is not an original idea, Ian says. "Historically, the forest has been a part of the farm. It may be more of a European tradition. We have struggled with this dimension of the imbalance of the farm. A seventy acre farm with fifty-eight acres of this kind of impenetrable jungle as the forest grew back up after having been logged. So we felt, maybe there's a gentler way of going about it with this land. Maybe the chain saw is not the answer."

A Farmland Trust

The farmland, still privately held, is in the process of being turned over to the Biodynamic Farmland Conservation Trust—a trust created to serve as a vehicle for supporting any initiatives which may arise in connection with the farm: research, education, apprentice programs, and so forth.

The trust was formed in 1988 because David and Claire Fortier wanted to ensure that the land would remain in agriculture after they pass on. The Fortiers dissolved the commercial farm corporation, Fortier Farm, and passed the title of their land to the Trust. As David explains, "a commercial farm was never what we really wanted. The only possible rewards with a commercial farm are economic, and the chances of that are slim hereabouts. We knew the farm needed the community aspect. To keep the land open requires a gesture of sacrifice from someone or some group of people who will donate land or otherwise make it possible. Land prices are just too high for it to happen otherwise."

As the Trust stabilizes, it will help support the operation of the farm in many ways, especially with long-range planning and financial support.

The Robbs and Fortiers look forward to the possibility of having the trust take on more responsibility so that others can donate land, equipment, materials, as well as financial assets. As biodynamic agriculture grows, so the responsibilities of the Trust could grow. Ideally the running costs of the farm should be met through the pledges of the families who belong to it, and fund-raising activity will happen in support of the trust, for capital purchases and capital expansion programs.

Set up along a standard model, The Biodynamic Farmland Conservation Trust is just starting to find its way. The trust is a non-profit, tax-exempt organization, so the land will no longer be subject to taxation. But the trustees are not sure they will take advantage of this opportunity. They may wish to go on paying taxes, as a good faith gesture to the larger community they are part of. However, they cannot afford to pay at the rate they are currently being taxed: as if the land were open to development as house lots. They hope to negotiate a rate realistically pegged to the land's value as open farmland.

One Step Further

Armand Ruby is one of the seven trustees of the Biodynamic Farmland Conservation Trust. A teacher of environmental science, one of his principal concerns is land usage.

In recent years he has studied many alternative forms of land usage and farming. But when he learned about biodynamics, he knew he'd found what he had been looking for. "The biodynamic view of the world takes the notion of Gaia—the earth as a living organism—one step further. If the modern view of Gaia is the earth as an interdependent super-organism regulating the earth's waters and atmosphere and so forth, then that view is expanded in biodynamics to place the earth in its cosmic setting: a living universe. So the concept of earth, or Gaia, becomes even broader or grander, deeper in meaning: part of a living universe."

"In the biodynamic view of the world, the earth's biota is far more than just a complex set of chemical reactions. The highly mechanistic view of the world as a set of chemical reactions is incomplete. There's more to it. There's a spiritual basis to everything that we see on the earth. When we see the universe as alive, we see that life is infused with cosmic spiritual influences, and we see humans as spiritual beings incarnated in human form. It's the same with a cow or a tomato. The physical part we see is not the sum total of the cow or the tomato. There's something more. And this is also true with the soil; it's not simply the constituents we can analyze in a lab with microscopes and so forth. In some way the soil, the earth itself, is also alive.

"Supporting a biodynamic farm does something more than just produce health-giving food for people; it also re-enlivens the earth, increasing its vitality. It is health-giving for the earth itself. Biodynamic agriculture is one of our best potential means for healing our relationship with the earth. We have done a lot to really damage the earth in recent centuries, and biodynamic agriculture is a way we can start to help heal that relationship. You can look at it from just an ecological level: how does it help to build soil as opposed to deplete soil; how does it help us to avoid polluting the soil and ground and surface waters. Then you can go from that ecological level to more spiritual levels as well. How does biodynamic farming help us to build our relationship with the earth as a living being?"

A graphic illustration of this respect for the land comes in a story Nicolette Robb tells. "If you go into the farmlands of this valley in the height of summer, there's not a bird singing. You can't hear one insect, the pesticides have been used so heavily. In the eight years that we have been here there's been a change in the wildlife around this farm. Birds and animals have returned in great numbers."

A Part of the World

"This is not a community that goes off into the mountains and builds walls around itself in isolation," Ian says. "Rather, it's a community within society, within an area that's just mainstream America, and yet these creative things have arisen.

"We want people to come out to the farm. We want to meet them, and to have them meet us and to see the farm." To abet member involvement, the farm has placed a large compost bin next to the parking area. That way, when community members come to pick up their produce, they can bring with them a bucket of household food scraps and dump it into the compost bin. They don't just take away, they also leave something.

At Brookfield Farm the emphasis of community support lies not so much in the direct relationship between money and produce, but more in the community's involvement in supporting the whole farm. As Nicolette puts it, "we seek to establish a common feeling based upon our interaction with people around the activities of the farm. That was the exciting thing to our first community members—not just a way of getting produce, but the concept of supporting the whole farm."

Naturally, Ian shares this feeling. "If you want to have people who are supporting an individual farm, or a group of farms that are working together, then it really is a case of supporting the farm and taking the good and bad times along with it. We hope there's good times," Ian says, "that the farm does well, and that the animals stay healthy, and that the land produces. But I get the feeling, and it would make me a little nervous if we were doing it here, that if we had a set fee for a set amount of produce, that people might wonder why their money was going for things that they don't get an immediate benefit from. They might say, 'Well, I'm just interested in vegetables, what are you talking about pledging for cows? And besides that, there weren't enough beets left when I came to the shop.' They will always be disappointed because you promised them two pounds of onions, and three pounds of lettuce every week, and what happens if you don't get that? You feel cheated. So we haven't let this happen. Of course, everybody's different and some will feel there's a direct connection between the amount of produce they receive and the pledge that they made. That's a reality. But most feel they are supporting the farm.

"Nicolette and I are not kidding ourselves that we have all the answers, but we do try, with the core group, to always bring our practical actions from the realm of the highest ideals for the land, the group, and the earth. So then you feel that you are doing something morally correct, that you are heading into the next century doing something that is really bringing the earth and humankind forward in the right way. If you really are coming from the highest principles, and you have all the other stuff that goes on, all the bookkeeping and all the finances and the nuts and bolts of the farm—they are important, they have their place, but it's the high ideals that sustain you. That's what gives us the strength to face all the changes and all the unknown factors. What if this year there's not a drop of rain? If it's a terrible year for farming, and at the end of the year there's no produce at all? Does that mean the end of the farm? Would people still be behind the farm if there wasn't much? What happens if disaster befalls us? That's why we're not just selling food, or a share in the harvest. That's why we want to involve people directly in supporting the farm.

"What can we feel secure about today? There's very little to feel absolutely secure about. The day has gone when I as a man in my thirties could look over a piece of land and feel absolutely secure for the rest of my life, and imagine that my children will be here to work on the land, that kind of generation after generation security of the land. That has already changed. I will be interested to see how Brookfield farm develops over the years. We want to steer it in a certain direction. We're not just open to any old whim, we have ideas. But because people are involved it's not just going to be mechanical. We have to take that into consideration. We have a plan. Every good farmer ought to have a thirty or forty year plan. You know, this is what you're working towards. In winter moments Nicolette and I can sit before the fire and we can see a park-like landscape, with animals foraging in the woodland. We can see it, but we have to realize it may never happen, or that it may happen in some other way. Who knows? We have to allow for that."

No Stock Formulas

"There's no stock formula," Nicolette says. "Ian and I have always felt a little inner reticence at being lumped under the concept of CSA because we feel it does an injustice to Brookfield Farm and its potential. There's so much more that can come from this farm. There's nothing wrong with the CSA concept. It's really exciting for the community to take initiative to control the way its food is grown. But here it came the other way. The farm needed to be cared for, and the future of the farm needed to be cared for. And it was from that vantage point that this group of people came together to embrace Brookfield Farm. There's a different emphasis."

"Some of the ideas we're working with are fine for a model," she says. "They can give you things to think about—issues of ownership and financing, and all the nuts and bolts. But we soon realized that there was no easy route. We couldn't pick up a paper describing some other farm and place it down over our land as if it were a model, as if you could stamp it out on an assembly line. And you couldn't pick up the model of our farm and place it over another seventy acres and say that it's going to go exactly the same. I doubt that."

Does the broad model they are establishing have potential for the future, or is it only for a sophisticated handful? Nicolette says "There are elements that can be duplicated. But you need to acknowledge the wider ramifications of this, that you're not just dealing with the economic structure of the farm, and its need for economic support. The relationship that can develop between the farm and its community allows for a mutual support system that encompasses far more than economics alone.

"If the farm is solely an economic model," Ian adds, "and you are attracting your support purely on an economic basis—it's cheaper or some such—then it's all in the realm of the economist, not the farmer.

Because all you would need then is for an even smarter economist to come along and think of an even smarter economic plan with slightly different variations. We can't think of everything. There's probably better ways of doing this. But we are receiving the support we need.

"We're all pioneers of this way of farming, of this kind of relationship to the land. But I'm sure we are the predecessors of the predecessors of the new type of farmer. There have to be many more farmers involved with this, and it will have to stand the tests of time."

The Farm Revisited

The tests of time took a toll on Brookfield Farm. Major change has come, but the farm endures as a CSA.

By 1994 Ian and Nicolette had parted ways, David and Claire Fortier had both passed away, and Brookfield Farm had gotten into financial difficulty. The operation was running a $15,000 deficit, and the cost of a share had climbed to $900 per household—high enough to erode the membership base. Things needed to change, swiftly and dramatically.

The transition came in 1995. The Board of Directors of the farm asked Dan Kaplan to manage the operation. At the time, Dan had seven years of growing experience. In 1993 he had worked with Trauger Groh at the Temple-Wilton Community Farm in New Hampshire to learn more. He and his wife, however, were ready to settle into a more permanent arrangement, and so when they heard that Brookfield Farm needed help they made the move. They were interested in a farming situation with stability, and in making the economics of farming on a modest scale work.

"What I saw," Dan says, "is that the farmers sacrifice a lot to keep a CSA going. They ask very little for themselves, work far too hard, and make sacrifices through the year to keep the CSA afloat. That's not a prescription for long-term success."

"What I did when I took over as manager of Brookfield Farm," he explained, "was to shrink the scope of operations. I tried to take off some small pieces that were burdensome, and make what remained very tight economically. I cut the number of animals. I kept the Dexter cows, and some of the chickens and sheep, but got my neighbor to grow our hay. In general, I just cut things to allow a greater degree of labor efficiency.

"We put the shovels and forks and other hand tools away, and changed the arrangement of the fields so that we could get a tractor into and out of them more easily, and really benefit from the mechanization of a tractor for the field work."

He also cut back on the amount of food that shareholders were receiving, and on the amount of money people were paying for a share (in 1996 a full share was not $900, but $450, and a half share was $250). "It was way too much food for the members, and way too much expense," he says. "That cutback was really needed. But we made sure people were getting a good value for their dollar. No matter what, no matter how high-minded the project, people always make the connection between what they are paying and what they are getting. That's just a basic thing that virtually every CSA member looks at.

"We made things clear and small—manageable for us, and for the CSA members. It had gotten too large and complex. Now members get less food, but pay less money and get more of the things they want to eat."

In response to the changes, membership has returned to a stable and supportive level: 300 shares in 1996, mostly local residents, but with 70 of the shares in Greater Boston, about ninety miles to the East. Dan Kaplan likes having the membership with some geographical diversification. While there are many places for people to get fresh organic produce in the Pioneer Valley, where the farm is located, there are relatively few places in Boston. The shares get snapped up in the Boston area, usually within a few weeks of being offered.

Organizationally, Dan Kaplan is an employee of the land trust, which owns the farm. He has two apprentices and hires seasonal labor. With the core group, he writes a plan for the farm each year, and then takes it to the Board of Directors for approval.

"I have a good life here at the farm," he observes. "I have an adequate salary, health insurance, and a manageable work load." The farm pays him a base of $15,000 a year, with incremental incentives based not on how well the crops do—since that is a risk that the whole farm membership takes together—but on how well he manages the business and how many shareholders he attracts and keeps. Without undue strain, but with careful management, Dan Kaplan has been able to average in the neighborhood of $30,000 a year in income, plus his health plan. And he gets time off in the winter to regenerate, and to plan carefully for the coming year.

Brookfield Farm has moved to Mix and Match (see *Example 10— Variations on a Theme*) to give shareholders a selection of the produce they want for themselves and their families. The farm also sells locally produced organic apples, bread, and milk. "Mix and Match totally works," Dan comments. "People self-select. It works much better than filling the shareholder bags with a ration of what we, the farmers, want to grow and deliver. People like it a lot better."

The farm hosts community gatherings about once a month through the growing season: planting and harvest celebrations, pot-lucks and barbecues. Usually about ten percent of the membership will turn out.

"Things have changed," Dan Kaplan concludes. "We've lost the romantic illusion of community, but gained a real community. Brookfield Farm is not seen anymore as something special and 'apart from the real world,' but rather just as part of the community, the way the local churches, or schools are part of the community. We are here. We are not the focus of the community, but we are a part of it."

Example 4

The Kimberton CSA

The Farm 1990*

In the hilly countryside of Chester County, Pennsylvania, astride the usually gentle waters of French Creek, the Seven Stars Farm has been a home for innovative experiments in agriculture for over fifty years.

Originally part of the A. Myrin estate, the land where the farm is located became a training center for biodynamic agriculture under the guidance of Dr. Ehrenfried Pfeiffer in the 1940s. With hundreds of people in America and abroad, Pfeiffer shared his experience as a farmer and technical advisor. Eventually, 400 acres of the farmland were donated to the Kimberton Waldorf School, which still holds title to the land. Now, on a small portion of that land, a CSA garden has taken root.

Just down the road from the school and the garden is the Biodynamic Association for farmers and gardeners who are interested in this approach to working with the land. Rod Shouldice, Director of the Association (in 1990), was instrumental in organizing the Kimberton CSA. He first heard about the idea from Jan Vander Tuin in 1984. Vander Tuin, who also inspired the Great Barrington CSA, was tremendously excited about the idea and wanted to make it happen immediately. But the time was not right. People just weren't sure the idea could work.

As Rod recalls, at least one organic grower experimented with subscription farming in the Pacific Northwest in the late 1970s. He created an arrangement whereby people came in the spring and placed orders for potatoes and carrots and so forth, agreeing to pay a fixed price when the harvest came in. But consumers were still buying their produce by the pound, not contributing directly to the support of the farm, as with the CSA concept, where the consumer agrees to take the risk with the farmer.

In the fall of 1986 the Seven Stars Farm was searching for something new. The school had been running the farm, and it had just not worked

*Reprinted from the original *Farms Of Tomorrow*

out well. Then Trauger Groh visited and gave a talk on the experience of
the Temple-Wilton community Farm, which had just completed its first
year. The Kimberton farmers were interested, but they felt they could
not just shift right over to a community supported farm. The farmers
were unsure they could find the right level of support because the scale
of their operation was so large: 400 acres, a scale which requires intensive capitalization to succeed. So after discussion they decided to start
with a small-scale project, a five-acre vegetable garden. Then people
could gain a gradual experience of what a CSA was like.

Here's the concept the community decided upon: they would form a
CSA (Community Supported Agriculture) where a group of families
would pledge together to cover the costs of the garden, including a
decent living for professional gardeners. Then, once or twice a week,
mature crops would be harvested and divided among the shareholders.
They would get a regular supply of fresh, healthful produce during the
growing season; and the growers would get an assured living. The farmers would not only be freed from the need to take upon themselves the
financial risks inherent to farming, but also from the need to go selling
in the midst of the growing season.

Shortly thereafter the Kimberton group contacted Kerry and Barbara
Sullivan, who they knew were looking for a situation, and encouraged
them to pack their bags. According to Rod Shouldice, this was the logical next step. "A crucial ingredient for a CSA to succeed is having gardeners who know what they are doing." In fact, one limiting factor for
CSAs is finding capable people who can grow excellent produce on a
large scale.

The group asked the Sullivans what it would cost for them to grow
vegetables that first year. The Sullivans said $24,000. Once an agree-
ment was struck, the Sullivans also advanced the CSA $10,000 of per-
sonal savings to buy the necessary equipment for a five-acre garden. The
loan is paid back to the Sullivans a little each year.

A Late Start

Kerry and Barbara Sullivan had attained a measure of fame as the coordinators of the North Carolina demonstration garden established by
Mother Earth News magazine in the 1970s. But after four years they had
journeyed to Emerson College in England to deepen their knowledge of
biodynamic agriculture. While they were away they hoped that there
would be a growing interest in biodynamic agriculture in North Carolina, and that they would be welcomed home for their expertise. But

when they returned there was nothing like that. They toured several bio-dynamic farms without finding a situation that seemed right for them. So they returned to North Carolina for a year.

But when the families of Kimberton decided to start a CSA and called, the Sullivans didn't hesitate. They arrived ready for work in February, 1987. When they arrived, there was no greenhouse, no tools, not even a spot for the garden, just a big hayfield in contour strips. The Sullivans set to work immediately, ordering seeds, buying a tractor, and securing all the necessary equipment. In southern Pennsylvania, farmers begin planting in March, so they had to rush.

The late start was a big handicap. Kerry and Barbara worked hard to build soil fertility and to keep weeds from strangling the thirty-three crops they had planted. But ultimately they brought in a fine harvest.

While the Sullivans were learning about the soil and the climate, the consumers also had to go through a period of adjustment. They had to learn to eat produce in season as it was harvested, instead of just picking by whim from endless choices in supermarkets.

The Mechanics

As Rod Shouldice recalls, "when we decided to start the CSA, families were willing to commit their money even though they had yet to meet the Sullivans. We walked out of our organizational meeting with checks totaling $15,000—that's serious money. Even though it was just part of the $24,000 we needed, it was enough for us to see that we could make it."

The first year the Kimberton CSA had sixty families. Then the phone started to ring. People had to be part of the project. "Once a CSA project has worked in the world for a year," Rod points out, "you are on solid ground."

They have decided to hold at 100 families for now, so this group of 100 families has to meet the costs of the CSA for the year. As with other CSAs, each year the core group meets and makes up a budget for the coming year based on the expected expenses for the coming year. Then they call a meeting for all the families. At the meeting the families pledge financial support for the CSA. The Sullivans have found it important to have people make a deposit or to pay in full early in the year so they will have cash on hand to buy seeds and supplies.

As it stands now, one share is approximately enough for a family of four, but there's such a range in eating habits that one cannot depend on that rule of thumb. Some families of two might eat a lot of vegetables and easily consume one share, whereas a family of five or six might eat few vegetables and so find one share sufficient. Some couples use more than a family of eight. Families who do extensive canning and freezing will invest in two shares.

Not everyone pays the same amount per share. As Rod Shouldice explained, "a childless two-career couple can afford to support the garden more generously than a single-parent family, and we wanted both to be able to be members. So when we had our February membership meeting, we told people they could pay from $270 to $370 for a share, as long as the total averaged out to $320. Everyone wrote down the support they could offer, then someone gathered up the papers, went in another room and tallied it all up. It turned out we were short an average of five dollars a share, so we passed the papers back out for another go-around. This time, we ended up with a surplus of ten dollars a share."

In 1988, for example, a share cost each family an average of $320. Because of the drought, vegetable prices in local supermarkets skyrocketed, and the cost of an equivalent amount of vegetables was $530. So members of the Kimberton CSA not only had the benefit of fresh, chemical-free local produce, they paid $210 less for it.

The Kimberton CSA wanted to create a way to share the harvest according to need. While the Temple-Wilton CSA in New Hampshire simply lets its members come and take what they want, that didn't feel right to the folks in Kimberton. Some people were worried that the early birds would get the best of the harvest. So the Kimberton CSA took a middle approach. A sign in the distribution shed on pickup day tells members how much of each crop they are entitled to take. If you don't want all of your share, you simply place it on a surplus table. Anybody can take what they want from that.

Beyond saving money, which may or may not happen, why else would anyone want to join a CSA? Barbara Sullivan responds by asking "have you been buying your vegetables from the market? Anyone who buys supermarket vegetables would have two darn good reasons to join a CSA: quality and freshness." Kerry adds that lots of people join after tasting a sample: "they visit neighbors who are members of the CSA. After they've had dinner and taste the food, there is no comparison. They have a hard time going back to the supermarket, where the food has very little flavor. Also, people know they are supporting something environmentally sound. Consumers support the CSA because they know that this approach to farming helps to heal the earth. And, they sense the difference in nutritional value. That the food had more vitality and life to it."

People also join because of the community impulse. It is a group of people doing something good, something aimed at a healthy future. As with many other CSAs to date, there is a waiting list of people who would like to join.

Pluses and Minuses

As they have worked within the CSA concept, the community in Kimberton has identified several distinct advantages and disadvantages.

According to Barbara, one big advantage is not having to market during the growing season. They put a significant amount of time into communicating with all the families and so forth, but they do it in the winter. They don't have to go out and sell while they are growing.

Another aspect that appeals to Barbara is that there is little waste. "You don't have to toss things away because they are not all a uniform size. But uniformity has nothing to do with quality. Our produce is generally the same size, but not carried to an extreme where every cucumber is exactly seven or eight inches long, and the rest are thrown away. Our produce looks great. It's all harvested and washed and displayed beautifully in the area where people come to pick up. But there's no rigid uniformity, and very little waste."

A third advantage is that the consumers know where the food came from, and the growers know where its going, and who is eating it. They talk with each other. Consumers tell the growers whether something is good or not, and they can complain directly to the growers if something is wrong. "When we were deciding what we wanted to do," Barbara says, "one thing I knew I didn't want to do was to be a market gardener and just send the food off somewhere. I wanted to feed people that I knew. I wanted to be more self-contained. It was and is important to me to have a direct connection with the people who eat the food."

Kerry also has strong feelings about this direct connection between consumers and growers. "People will tell us, 'you know I could never get little Johnny to eat beets, but now that he knows they are from the garden he's happy to eat them.' Little things like that are important to us."

Barbara says "with the CSA concept people don't just look at their families and try to get the best deal for their family. They look at a community of one hundred or so and try to do right by the whole community. We are all trying to eat from the same garden. We don't all make the same amount of money, but then we don't all pay the same thing for our share in the garden. One person will pay more out of their free will because they know they have the ability to do so, and another person will pay less because they don't have it."

Kerry and Barbara agree that knowing they have a set income is a real advantage. The consumers take the risk with the growers, even if the crop fails completely. "Last year we couldn't irrigate our potatoes," Kerry explains, "so they failed in the drought. We didn't have enough equipment to water everything, so we just had to let them go. We had to make a decision. But the whole CSA community takes the risk and so takes the loss. We still had enough income to go on with our lives without getting into debt."

While the Sullivans may not have gone into debt, their income is low. In 1987 Kerry and Barbara earned about $16,000 (not including $2,700 from the CSA in repayment of a loan). While that's a low wage for two people who each put in about fifty-five hours a week of hard work, they reason that the CSA is in its infancy and that like any commercial enterprise it may take a while to begin paying off. Also, they are well aware that if they had become market gardeners they would likely have worked harder—spending many hours selling the produce—and most likely would have ended up with less income. And for all that, they would not have had the satisfaction of being involved with a new concept that re-knits the community together and is gentle on the earth.

Another disadvantage according to Kerry is that it's fairly hard to mechanize to the extent a big commercial operation would, because you have so many different vegetables. "Instead of being a market garden where you choose four or five crops that you can do really well, you have to grow many different things. We grow over fifty different things—so its like a huge family garden, rather than a market garden where you have a half acre of peppers, and an acre of cantaloupes." But Barbara adds that while this is a disadvantage from one point of view, its an advantage from another. "It's what we were looking for. We'd be bored with just five crops. And also we are interested in training other people. This gives us all much more to study and learn from, a much better training situation."

Exploring Biodynamics

The Sullivans began to learn about gardening in-depth when they apprenticed in California with Alan Chadwick, the man who brought French Intensive gardening to America. Then they went to North Carolina in 1978 and founded a demonstration garden for *Mother Earth News*. The magazine was starting EcoVillage, a 600-acre site demonstrating some of the ideas explored in the pages of the magazine: solar architecture, organic gardening, and so forth. While in North Carolina, the Sullivans lived in a yurt, gave demonstrations and workshops, and took on apprentices. From Alan Chadwick they had learned much, but it was at *Mother Earth News* garden that the Sullivans really became interested in biodynamic agriculture.

With advice and encouragement from Peter Escher, a biodynamic consultant, they began to experiment with some of the biodynamic practices. Then in 1983 they went to Emerson College in England to study biodynamics in depth. Why go to England? According to Barbara, "Emerson College was the only English-speaking place we knew of that offered a strong theoretical grounding in biodynamics as well as practical training."

"When we went to Emerson we didn't necessarily learn a bunch of facts and formulas," Barbara says. "It's more like learning a new way of thinking so you can solve your own problems. We don't just go to a textbook and open it up and see what a solution might be. After all, nature just doesn't work that way. You can't apply the same solution twice in the same way. It doesn't work.

"At Emerson we did not learn the chemical kind of thinking where you take this much out of the soil and you have to put that much back in, so-called quantitative or mechanistic thinking. But what we learned is much broader. It takes in factors other than the chemical processes that go on. It's an approach that tries to understand the life of the soil. The soil is not just dead minerals. That's important to be aware of if you are working with the soil to coax life out of it.

"We didn't go to Emerson because we had heard ideas and thought, hey, that's a great idea. We went because we tried the biodynamic preparations and approaches in the garden and they worked. So we went for very practical reasons. We saw the results first and we wanted to know why, why does this work?"

Kerry offers an example of this from their direct experience. "We could see the difference in making compost with the biodynamic preparations, and making it without. At that time we were working on a very intensive scale in North Carolina, where we could see results very easily. Here, where we have ten acres to take care of, we can't look at things as closely. But when you work a half-acre with a spade and fork you see a lot more. So we noticed things using preparations in the greenhouse, in the compost, and on the plants. It made a big difference. The compost broke down a lot faster. But what we were mainly impressed with was how thoroughly it broke down, and how much better a product it was in the end. It smelled a lot better. In the greenhouse, seedlings rooted much more thickly. And it was just better all around."

Rod Shouldice, the Director of the Biodynamic Association (in 1990), adds that "rather than just trying to feed plants, biodynamics tries to make the life in the earth stronger and more vital. The earth is made sensitive so it can give the plants what they need. It's a crucial difference. The day of artificial fertilizers and pesticides is over. It's dying now."

Gardening Techniques

The Sullivans employ a number of innovative techniques to grow vegetables for 100 families. They prepare, fertilize, shape, plant, and weed hundreds of permanent raised beds mechanically, using methods they learned from biodynamic grower Mac Mead in Spring Valley, NY.

They have a 1950 Farmall Super-A tractor with well-spaced front wheels so it can straddle their four-foot wide growing beds. To make the planting beds, they hook up a disc harrow behind the tractor and then later make passes with a specially built bedmaker. The bedmaker, designed by W.W. Manufacturing Co. (60 Rosenhayn Ave., Bridgeton, NJ 08302), has one large disk on each side to push dirt to the middle, adjustable boards to shape and level the soil, and two rows of small, soil-pulverizing disks (called Meeker harrows) to do the final texturizing. All they need to do is to make a couple of passes with this rig and a bed is worked up for planting.

Then they pull a homemade row marker down the bed, and it makes three precise planting furrows. Finally, they push a Swedish Nibex seeder down the rows by hand. The seeder drops accurately spaced seeds in place. Seedlings, though, are planted by hand.

When it comes to weeds, the Sullivans use a special-ordered Buddingh Wheel Hoe (Buddingh Weeder Co., 7015 Hammand, Dutton, MI 49316). The hoe has a double set of gear-spun wire baskets, and it fits the row spacings in their beds exactly. When they pull it down the rows the baskets churn through the soil and uproot the weeds. Any weeds not reached by this contraption are removed with hand-held scuffle hoes.

In the greenhouse, the Sullivans make use of Speedling seed-starter trays, while out in the garden they use drip irrigation to conserve water, and floating row covers to protect tender young plants from frost and insects.

Seven Stars Farm

The Kimberton CSA is now nestled into a corner of the 400-acre Seven Stars Farm, but the garden and the farm are not completely distinct organisms.

For a long time the Kimberton Waldorf School tried to run the farm by hiring farmers. But that arrangement was generally unsatisfactory. So in 1988 they decided to lease the farm to David and Edie Griffiths, who agreed to farm it biodynamically.

In 1989, with the other people who help farm the land, David and Edie Griffiths began the process of making a long-term commitment to the land, entering into a twenty-nine-year lease with the school. A committee from the school will oversee the lease with the farm. The farmers will oversee the land and maintain the buildings.

The Griffiths plan to form a cooperative corporation of all the people who work on the farm or in the farm store, and then to assign the lease to the corporation, modeled after the Spanish Mondragon system. This, David says, flies in the face of many of Trauger Groh's ideas, because it concentrates all the responsibility on the farmers and workers, rather than spreading the risk throughout the community. But after much deliberation, the Seven Stars farmers feel this new form is a good thing, and that it will work well for them.

With this cooperative approach, the management of the farm is creating a corporation that will live on beyond the individual people who are involved. The leasing arrangement will solve the problem for them of binding up a huge amount of equity in a major banking risk.

The school, meanwhile, is attempting to sell the development rights to the land to the state under a new Pennsylvania law. That will ensure that the land is always used for agricultural purposes, a real concern as development continues to spread out from Philadelphia.

As for the CSA garden and how it fits in, David Griffiths describes it as a symbiotic relationship. The CSA makes an agreement with Seven Stars Farm to lease the acreage for the garden—five acres in 1988, ten acres in 1989. Beyond that, the particulars of the relationship are still in process of being worked out.

David is an enthusiastic supporter of the CSA concept. "I must stress the importance of a CSA for a large-scale farm operation. Everyone talks about farm preservation, but this is something concrete. It brings people into direct relation with the farm and the farmer when they come to get a box of food. There's real contact with the farm, and that builds support for the farm as a whole. It's something real, not an abstraction you read about in the paper."

Sales at the Seven Stars Farm Store have jumped nearly twenty-five percent since the CSA began. When people come out to pick up their vegetables, they naturally stop into the store to pick up other supplies. So the CSA can help anchor a large, non-diversified farm by forming a support network.

Both the CSA and the farm are doing the best they can separately. So at the moment joining together in some more deliberate way seems to be an unnecessary complication. But that may happen later down the road.

The Sullivans get everything the farm gets for the same price the farm pays. They use tractors or other heavy equipment for an hourly fee, and pay a small sum for manure and compost from the farm. It's a business arrangement. But as David points out, "we can't let business hide the fact that we are all farming the same farm. It's not our manure, or their manure. It's the farm's manure."

Realistic Idealism

Kerry and Barbara Sullivan intend to stay with the Kimberton CSA over the long term. Barbara says "in some ways our scope is more limited, because we are not trying to create a whole farm organism within the CSA, because we are part of a dairy farm and that's where we get our manure. We don't need to get animals. It's a real difference from an operation where the CSA is the whole farm. Our CSA is just the vegetable garden within this farm. That makes it much easier on us."

The Sullivans have no overwhelming urge to expand. As they see it, there's a whole world in seven acres—lots of room for refinement of what they do, and improvement in the CSA. "Within what we are doing now, we could take a major step to growing most of our own seeds. That's a big step we aren't ready for just yet," Barbara says.

The CSA concept is catching on, but as each group discovers, there's no formula. Because the people are different and the circumstances they face are varied, each CSA does it its own way. As Rod Shouldice says, "Every project is different, and should be different. One CSA can't be reproduced elsewhere. The group that's forming it must be realistic about where they are at. I encourage them to stretch their idealism—but still they must be realistic."

Rod believes the CSA movement will spread widely. "I sense a continued strong interest. People frequently call the Biodynamic Association for advice (the Biodynamic Association has a toll-free phone number for inquiries; see *Appendix F—Resources*). But the CSA is not a finished product. It's not perfect, but at least we feel we are heading the right way. If we can resolve the problems, then CSAs are definitely here to stay."

"It's easy to imagine vacant city lots bursting with organic vegetables, or small groups of city dwellers getting together to hire a gardener to produce vegetables in the suburbs for them. It could even work for a corporation to hire gardeners for a six-acre plot behind the corporate center, and then saying to employees, 'Besides picking up your pay check once a week, we're going to give you a basket of vegetables, too.'"

The media is very responsive to the idea of community supported agriculture, according to Rod, so a core group can take advantage of this to help them get started. He recommends first having a public lecture to get people aware of the idea, then a follow-up meeting, and possibly a story in the local newspaper or radio station. "By then," he says, "you'll know if there are enough families willing to put their money on the line."

The Farm Revisited

"Things are going well," say Kerry and Barbara Sullivan. "A lot has changed, but on the whole we find we are doing well. We are rooted now. You can feel it. Our CSA is something people really need and want. And that makes us feel good."

As they see it in the autumn of 1996, their CSA garden gets more substantial and more spirited each year. They have made hundreds of improvements and refinements over their ten years of community stewardship. The garden is now fully certified as biodynamic by the Demeter Association, and it routinely feeds and nurtures a large community of 150 households.

"As a gardener," Kerry explains, "you always have a picture you are moving toward for the garden, what it will be eventually. Our garden is making progress. It's not fully 'there' yet, but so much more of it is there."

"The soil is stronger now," Barbara says, "and as a farmer you really feel how much the soil has improved. There's more vitality out there." Kerry gives voice to their shared observation that there's also "something noticeable where, when you stand out in the garden, you feel uplifted. A kind of spiritual atmosphere develops over time, over the years."

"We are fortunate to have people come to the garden to pick up their food," Kerry says. "We don't deliver anywhere. The people have to come here. As a result, they have an actual connection to the land, and that has made a big difference. They make a connection not only with the farmer and the produce itself, but also with the garden itself—the place where their food is grown."

"Another key," Barbara explained, "is having people who have been part of the CSA for ten years, for ten seasons. That's rewarding. There is just a certain core of people who become the farm family, because they are here all the time over five, or six, or ten years."

The money part also works now for the Sullivans. For the first three years, they did not make enough to cover all of their living expenses. But they worked that out. As Kerry explained it, "we finally decided we had to ask something reasonable for the vegetables. We were asking way too little money for all the work that needed to be done. We were getting burned out, and we couldn't cover our expenses. We finally realized that what we were doing, and the food that we were offering, were really different. You cannot get the quality of food at the supermarket that you can get here. It's worth more. We had to ask for more, and we finally felt comfortable about asking for a substantial increase in the cost of a share. When we did that we lost some people. But that was alright. We needed to cut down on the size of the membership based on what the land, and what we as the farmers, were capable of doing. But I want to emphasize that the value is there in the garden and in what we produce. People are paying a fair price for the food."

"After we raised the cost of a share about four years ago," Barbara adds, "the membership had a whole different feeling. We ended up with the serious, committed people who really appreciated the garden for what it is. We lost the people who were just looking for cheap vegetables. The whole community has had a really different feeling since then. It was the best move we ever made. We could not have survived until we made that decision. You can start out the way we did, but sooner or later you've gotta have enough people who are really committed to making the CSA work. It's a real community garden now. We could go, and the Kimberton CSA Garden would still go on. This is a real people garden."

To anyone starting a CSA, they offer some straightforward advice: "Make sure not to price too low. Ask enough, but not too much. There has to be a connection between what people are paying, and what they are getting."

Another thing that has changed is that now there is a lot of "U-pick" for the Kimberton CSA members. As a consequence, throughout the season there are lots of people in the garden to gather peas, beans and other crops that are otherwise very time consuming for the farmers to harvest. "The U-pick actually adds to the value of the community farm," Kerry says. "People like it, especially people who have kids. It is really something to watch a family regularly go out and pick their food in a clean and beautiful place."

As of 1996, four people are working the ten-acre Kimberton CSA garden: Kerry, the master gardener, two gardeners who have a couple of years experience, and one apprentice. As for the level of work in the garden today, Kerry observes "it gets done. It never ends, but it gets done. I suppose it's that way on every farm."

About two years ago Barbara stopped working in the field. Her migraine headaches from the heat just kept getting worse and worse. "My first choice would have been to stay out there working in the garden, because that's what I love," she says. "But my destiny led me in another direction." She is completing her Agricultural Science degree at Penn State. Then she will perhaps get a Master's Degree in Alternative Agriculture. She is building her knowledge base and credentials, for someday she and Kerry would like to establish a school for biodynamic farming.

Example 5

The Hawthorne Valley Farm

The Farm 1990*

A rangy man of medium height, Christoph Meir has a farmer's strength in both his arms and in his convictions. He takes his rest gratefully, for the moments of repose are few and far between. Since 1974 Christoph has been director of the Hawthorne Valley Farm, an innovative enterprise located in Columbia County, about 100 miles up the Hudson River Valley from New York City.

The Hawthorne Valley Farm embraces 350 acres of rolling fields and forest, land rich with streams and ponds. This beautiful valley is the stage upon which a sizable community of people have joined forces to integrate agriculture not only with education but also with art. Hawthorne Valley stands alone, different from the typical CSA. The farm operates under the umbrella of the Rudolf Steiner Educational and Farming Association, a non-profit, tax-exempt organization, which also includes a school, a program for visiting students, and a number of artists and artisans.

As stated in the Association's brochures, Hawthorne Valley has a broad vision: "The age in which we live demands new capacities in all human endeavors: in the sciences, arts, humanities. These capacities are insight, purpose, practicality—the necessary foundation for action and leadership. At Hawthorne Valley the interaction of education, agriculture and the arts is intended to provide the basis on which such new abilities may be fostered." Since its founding, the Association has made notable strides toward realizing its vision.

How the Farm Came to Be

In the early 1970s Hawthorne Valley was primarily a visiting farm for educating young people about agriculture, giving them first-hand exposure to life and work on a farm. Then, as now, students and teachers

*Reprinted from the original *Farms Of Tomorrow*   **157**

worked together during a week-long stay, sharing the daily farm chores, seasonal projects, hiking, crafts and nature studies.

But with this limited focus, the operation was not prospering. The people involved found it difficult to run both a school and a farm, so they sought a farmer to take full responsibility for the land. Ultimately, they found Christoph Meir.

Christoph was farming near Geneva, Switzerland when he heard about the opportunity at Hawthorne Valley. The association had the farmland and the buildings, a few pieces of equipment, and $40,000 to fund the first year of operation. With this skimpy purse, Christoph had to buy the seeds, the cattle, and the equipment, hire help, and pay his own salary over the first year.

As he began, Christoph insisted that the farm be primarily just that, a farm, not a school farm. While students would be welcome, the main business of the farm would have to be farming. The visiting programs would have to be run by others whose main business is educating young people. When that agreement had been established, Christoph joined with other workers and began to build what is today a modern, full-scale diversified farm operation.

Virtually self-sufficient, the Hawthorne Valley Farm relies on its own feeds, manure, and compost. No chemical fertilizers, pesticides, or herbicides are used. Instead the farmers follow biodynamic principles to produce milk, eggs, grain, vegetables, meat, and honey.

According to Christoph, a typical year's planting includes 30 acres of wheat and rye, 30 acres of oats, peas, and barley, 30 acres of corn, and 110 acres of mixed grass and clover to feed the cows, horses, pigs, geese and laying hens. Five acres of the farm is planted intensively with vegetables.

In 1988 the farmers completed building a new dairy processing plant, where they produce cheese and yogurt from the milk of their seventy-five cows. The farm also has a bakery, which specializes in breads baked exclusively with the farm's own stone-milled wheat and rye flours. Most of what is grown and processed on the farm is sold directly through a farm store which has become so highly regarded that people drive miles out of their way to shop there.

The farm's gross sales are about $750,000 a year, which includes about $270,000 from what is produced and processed right on the farm. The rest comes in from retail sales and the other dimensions of the diversified farm operation.

Alternative Business

Gary Lamb is the Associate Director of the farm, and the manager of the farm store. He says that when he first came the farmers were selling milk, cheese, and bread. They had a little door on the edge of the barn through which people could reach to put their money in a cigar box and to pick up what they wanted. But after business picked up the farmers got tired of having to deal with the cash box and supplying the milk and everything, so they decided to hire someone to take care of it. To justify paying someone, it became necessary to add other products from other farms. One thing led to another, and the Hawthorne Valley farm store developed from a cigar box in a corner of the barn to a modern retail store with five employees and $400,000 in sales each year. As Gary Lamb sees it, "the store has grown from supporting itself and the farm, to supporting alternative agriculture in general."

The farm store sells bread, cheese, yogurt, and a wide variety of health foods, fresh produce, cosmetics, books, and crafts. The store also sells the farm's products by mail order and at markets in New York City.

Community Support

Hawthorne Valley has come a long way since the 1970s. But while the farm is part of a larger association of a school, artisans, and the staff of the visiting students program, it has not yet found the perfect mix.

As Christoph explains it, "community support is our weakest part. Before we got the retail store we had kind of a food coop, which was outside of the farm, and also a little farm stand where people helped themselves and just left money in a box. We tried to have a farm pro-ducers-consumers association, and we had some meetings to discuss prices, but we realized that we farmers had to organize all the meetings. If we didn't push it, it didn't happen. So, we talked and said 'if the only interest of the consumers is to get cheap products, then forget it. We are busting our backs and not making ends meet.' So, we went more toward the retail. That was in 1979 and it marked a new beginning here.

"We felt there was no community support at all, except for some of the older friends who had struggled with us in the beginning. And we felt we were being taken for granted. People were grumbling, 'why is our food so expensive? Why does it cost more than the supermarket?' A lot of people either had their own food coops going or they just went to the supermarket—members of our community, people who had children in our school, would do a little token shopping here. It came so far that we went to a meeting and told people that if they did not change we would have to give up. We send our children to the Waldorf School at the farm and make a big sacrifice; we pay for that school out of our farming. It costs the farm about $20,000 a year. That's a considerable financial burden. Most farms don't have that kind of expense." While this is true, most farms pay taxes; as part of a tax-exempt organization, the Hawthorne Valley Farm does not.

Finally, Christoph explains, the farmers went to a meeting in 1987 where they announced that since they could no longer afford to send their children to the community Waldorf school, they were going to start their own school. In the intense discussions which followed, Christoph says, the whole community realized that the farm, as well as the school, required their wholehearted economic and moral support.

An Associative Economy

One of the strongest features of the Hawthorne Valley Farm is the economic association among farmers, processors, retailers, and eventually consumers—though the consumers haven't yet been fully involved because they are not organized.

Christoph explains the farm's concept of association by referring to the work of Austrian philosopher Rudolf Steiner. "He gave an alternative to communism and capitalism. He indicated that what you have to do is that you must consciously separate an economic sector such as producer, processor, or distributor, and that they should be organized on associative principles. . . . Hawthorne Valley is trying to apply certain of these principles, as far as we can—but we can't do it all in the existing social-cultural context. We can do it small, as sort of a model.

"Early on we realized that we have to develop associations where we discuss prices, where we discuss the needs between farmers, processors, and retailers. You just realize that as a farmer you are always at the short end of the stick. You are depending on what they are willing to give you. The retailer says 'well, I need so much for my living. I have to have so much of a markup,' and the farmers just cannot do that. So we decided to integrate the various facets and become our own processor and retailer. Now we have these seven businesses together under one roof: a farm operation, a milk-processing plant, a bakery, a retail store, farmer's markets in the city, a mail-order operation, and a tractor dealership that sells about twelve tractors a year.

"Now the prices I need as a farmer for my milk, not speaking now as the director of the whole operation, but just as the farmer, I need fourteen dollars per hundred weight, otherwise I cannot make it. And the cheesemaker will say 'yes, but if I have to pay so much for my milk, I really have to charge more for my cheese and my yogurt.' And so we talk again with the storekeeper or wholesaler, who is tied together with us under the association, and he has to sit and listen and we have then to understand and agree with one another while the other one meets that margin or price, and whatever quality questions there are. And we have to come to consensus.

"The only thing we are missing, really, is the consumer representative of our association, who would then sit together with us and explain the consumer's point of view and tell us why prices are good or bad, or comment on the quality. The customer and the producer then can negotiate and maybe come to a compromise."

The consumers directly support the farm, but in a free way. For example, they help to finance the farm whenever capital is needed for development or expansions. A few years back when Hawthorne Valley wanted to build the farm store, they sent out a letter requesting loans. They offered what the money market was offering at the time, about eight percent interest. They were able to raise nearly $300,000, and to get the money for a much lower cost than they would have to have paid to a bank; meanwhile, the community members were still getting a reasonable return on their investments. Additionally, they had the satisfaction of knowing exactly where their money was and what it was doing. When the farm does borrow from community members, the loans are for three or five years, and lenders have the right to recall their loans with just three months notice. The farm issues quarterly statements so everyone can see how they are doing with their business and in repaying the loans, and the books are open so any community member can come in and see what is going on. All of this helps to create a strong bond between the farmers and the consumers.

The Farm Workers

The Hawthorne Valley Farm provides enough for its fifteen farm workers to make a living: about $1,000 a month in pay, health insurance, room and board, and tuition for their children at the Waldorf School, a top-notch private school that otherwise would cost about $3,000 per child each year.

Yet the workers build no equity in the farm, or the houses where they reside. Technically, they are all employees. According to Christoph, they don't want any ownership or equity in the farm. "I strongly feel," he says, "based on Rudolf Steiner's indications on the three-fold social order, that farmers should have access to the means of production to use as we see fit as long as we can work them. If we don't produce, we should pass them along to somebody else who is able to use them."

But as it stands now, when workers at the Hawthorne Valley Farm retire they have no pension or other means of support. This is one of the many issues that the farm workers are grappling with now, to find a way to provide for their later years. At present the economics of the farm make a pension program impossible. Christoph feels the only sensible way to meet this need is to improve the economy of the farm: "You should be paid enough for your products so that you can make a decent wage and so that you can also provide for your old age. It shouldn't be any different than any other business. We have raised four children here. We might not have been able to put any money aside for our old age, but our children have been given good private-school educations."

Christoph says the farm would also like to involve more apprentices. To date, in fact, the development of the farm has been possible only because of the dedication of the apprentices, who earn even less than the workers but who are gradually trained in the sophisticated techniques of biodynamic farming. Hawthorne Valley intends to develop a comprehensive training program, including classroom work in the winter, for about twenty apprentices.

The Question of Land

While at Hawthorne Valley the farmers do not have ownership; the owner, the Rudolf Steiner Educational and Farming Association, has granted them the right to work the land and to use the products of the land as they see fit for as long as they are capable of doing it. Thus, at the end of their work, the farmers will not be able to sell the land for their own gain—the way farm families have typically financed their retirement. The Association holds title to the land and the property for the benefit of the community and to ensure that it is used productively and preserved for agriculture.

Just knowing that the well-intentioned Rudolf Steiner Association owns the land is insufficient security for the farmers, so they have approached the American Farmland Trust. With the Association, they intend to deed the development rights for the farm to the Trust so that the land can never be subdivided or sold for other purposes. That way if the Association were ever to go bankrupt and the land to come onto the market, there's a possibility that another farmer could buy the land and continue farming here. Real estate values are so high now—about

$10,000 an acre in the region—that nobody could ever farm the land at that rate.

Christoph feels strongly that in general, for all farms, land must cease to be a commodity. Either the farm must give away or sell the development rights, or find some other way to remove the land from the pressures of the open market. "Today farmers have to deal with buying land and making mortgage payments, but that should not be a concern of the farmer. Land for the farms of the future will have to be provided by the community at large, or in some other way, to relieve the farmer of the burden of paying for land in a situation that's really not economic. Land prices have no relation to reality or incomes, and the farmer is the one who feels this the most. That problem must be solved."

In contrast to private ownership, Hawthorne Valley's land is held in trust for the community so that no individual really has direct ownership of the land. The community supervises to see that those who are using the land are using it appropriately.

Healthy Experiments

Reflecting on the experience of his own farm and the growing CSA movement in general, Christoph says "I consider what is happening at the Temple-Wilton farm to be a healthy experiment. It's important for a community of people to be working together to understand again what farming is about. But I don't think it is the wave of the future. I think instead we are pushing into an associative economy, which is not going out to make as much money as you can, each one for himself, but rather to understand why we have to provide each other with the means of income, so I can produce for you and you can produce for me."

Christoph regards the Temple-Wilton Community Farm as an experimental form which can work in certain situations. But, he adds, "I don't think that it's a solution for a large population. Temple-Wilton is very important, but for us it's not the right way. We want to push more into something which can lead to widespread use—an associative economy, which means, in a way, price fixing, excluding competition. The producer will tell the consumer, 'this is what I need to produce a bushel of wheat,' and not a competition to see who can make the cheapest wheat on the market, and that will be the one who makes it while the other farmer goes bankrupt.

"What's wrong with setting it up so the one who survives is the one who can produce the cheapest wheat? That principle can work in manufacturing or with industrial goods," Christoph says, "but what happens when you apply it to a living situation like health and hospital care? You can provide cheaper health care by sending a robot around to the beds to make them up and bring tea to the patients. Yes, it's much more efficient. You can build a big hospital, and you can have a computer screen where the doctor sits and decides who needs what. But if you apply the same industrial-economic thinking which is fine for building cars, if you apply that to life, to growing food, to working with nature, or to health care or education, it's not going to work. It brings destruction.

"So the fellow who provides cheap eggs, what is he going to do? What has he done in the past? One fellow figured out that if he controls the environment of the chickens—lights, temperature, air, puts five or six in a little cage on a conveyor belt—it's very efficient. You can produce eggs much cheaper. But what are we doing to the animals? What is the quality of the product? It's cheap, but is it really cheap to eat those eggs? You haven't figured in your health costs, all the diseases and mental disabilities which you might get from eating crappy food. It costs in the long run. You have to include those costs also.

"What we need, really, is a different way of thinking, a different approach. You see our big problem is not farming. We know how to farm organically and biodynamically. We can improve a lot there, but our problem is the social and economic structure in the world. Because if we continue this way we are going to destroy our forests, our waters, the ozone layer, and our children with our school systems. We destroy our social relationships through the health industry. I mean, just look and listen to the words applied to it. It's incredible. I mean how can you have a 'health industry,' that's considered an 'investment opportunity' for people to make large profits from illnesses? We need new thinking here."

With conviction, Christoph says experiments like the Hawthorne Valley Farm, and the Temple-Wilton Community Farm are essential specifically because they help to test new thinking and new models.

The Need for New Thinking

Christoph raises many provocative questions about farm economics. "What are you forcing your fellow man to do when you make economic choices at the market? Are you forcing him to spray his land with herbicides and pesticides because he cannot afford to work organically, because you are not willing to pay for it? Do you force your fellow man down in Brazil or Nicaragua to spray and chop down his woods and plant coffee for you because you're not willing to pay the price of clean food and coffee?

"We are seeing a change now," he says. "Big businesses are jumping into organics. They realize that consumer perceptions are changing about poisons in the food. So Kraft, Nabisco, and all the rest are going to move into it. But if you don't change your thinking, and you're still into the corporate way of money making and profit maximizing and industrial efficiency, then you really change nothing. You're going to grow food a little bit healthier, but you're still not talking about self-sufficient farm organisms which can produce their own fertilizer. You're actually only adjusting your big monocultural approach to organic—starting to import all kinds of organic fertilizers—your way of thinking and your procedures are not going to change.

"If we integrated farms with a vegetable operation where we grow thirty different kinds of vegetables for a small market, we can never compete with the fellow who grows thirty acres of carrots only, and he specializes in that. He has one kind of equipment, one set up. Why is that a bad thing? Because," Christoph says, "such a farm has no diversity. It relies on monoculture with no crop rotation. The farm will always have pest problems which the farmer will have to deal with using pesticides, whether they are synthetic or organic."

The Need for a Broader Context

The Assistant Director of the farm, and manager of the retail store, Gary Lamb came to Hawthorne Valley in 1985. He was looking for a business position, and he wanted his children to have an opportunity to attend a Waldorf school. He was also concerned about a broad range of social problems, and he saw agriculture as an important and fundamental dimension of social culture.

As with Christoph, Gary has reservations about the typical CSA concept. "The problem with community supported agriculture," he says, "is that you have high concentrations of people in the cities and then these CSAs out in the country. Well, where are the people in the cities going to get their food? How can they work in a constructive way with agriculture? There has to be an association of food consumers in some way.

"The CSA is a method by which the farmer is assured a market, and that takes care of one of his principal worries: somebody to sell to, and a fair price for his products. Also the farmer has money in hand to invest in the land, and that, of course, is another worry. It's certainly better from the farmer's perspective than taking a loan at the bank and then having to pay interest on the money for this year's crop. But the real question is how do you expand this? It works on a local level for the local farm, but we have to find a way to make it applicable to agriculture in general.

"Here in the Northeast the winters are long and the growing season is short. People can support a farm in a region, but what about the farmer who grows oranges for you in Florida? Or lettuce in February in Arizona? How can you support these other farms, too? People are going to eat bananas and oranges and other things out of season. Community supported agriculture does address many key issues on a local level, but the percentage of agriculture that it can deal with, at least as far as the form is presently conceived, is limited. I think there's ways that can be found to get beyond this. And we must. We can't be selfish and just support our local farms; we have to broaden our response to other farmers. . . . In terms of the farms of the future, whatever we develop has to embrace all of agriculture. We need a broader context than at present or offered by the CSA concept as presently formulated."

Based on his experience, Gary believes that in the future there has to be a place for the checkout counter within the CSA concept. He sees the checkout counter and the choice that it implies as good ideas. Some people involved with CSAs think stores and their checkout counters are dispensable elements in getting food from farmers to consumers. But Gary believes they may not be. At Hawthorne Valley, the retail operation not only supports this farm, but also dozens of other farms and producers. "When you have a CSA," Gary says, "it supports one specific farm. There is a place for a store that supports a number of farms and a wider variety of goods."

Gary has some pointed advice for people who are attracted to farming as a way of life: "Get away from wholesale and focus on retail sales. That puts you in direct contact with consumers, which is critical. That's a beginning, to create some kind of understanding, to give the consumer and the producer a chance to enter into a dialogue. From there the relationship might evolve, the consumer might have some land, or some money they want to invest; or they might come to support you more directly, as with the CSA movement."

Get the Government Out

When it comes to the role of government in agriculture, Gary Lamb has an unyielding point of view. No doubt his position crystallized in 1988 and 1989 when New York State began thinking about banning raw milk. Hawthorne Valley Farm is the largest producer of raw milk in a state where there are only about seven producers altogether.

When the move toward banning raw milk intensified, the farm saw it as an opportunity to look again at the relationship of the government to the farm. Gary says that, ultimately, all agriculture is community supported unless the government is involved. "Many agricultural problems are due to government intervention, especially in the 1980s and now in the Midwest, where you have all of these farms going out of business. They have been taking out huge loans, and have been directed by the government to use chemicals in various ways, and subsidized to grow things that people don't really want or need. The enthusiasm, or connection to agriculture has been diminished. If you are growing a crop that you know people aren't going to eat, that is just going into storage somewhere or that will be dumped out so that you can get paid a subsidy, then there's no real connection to the earth. No human being will develop a connection to the land or the soil and go to all the effort it takes to grow food and then just watch it be thrown away. Government intervention introduces elements that have nothing to do with agriculture.

"Government ought not to be involved in agriculture at all. If there's going to be a Department of Agriculture then it ought to be run by farmers. After all, what does a politician really know about agriculture? Safety and other issues may be appropriate for government regulation, but when you are talking about technical expertise with crops, or the relationship of the farmer to the consumer, then history clearly points out that the government has botched it up. This is a crucial issue. Historically, if you look at the worst situations that farmers have had to face, you'll see that they are somewhere connected with government involvement and 'support.'

"Let's assume the farm of the future is going to be a low-input, non-chemical organic or biodynamic farm. What has already happened in the organic and biodynamic circles is that they've gone and been supportive, or are even pushing to get government standards for organic farming. That's the last thing to do. There's enough history to prove that getting mixed up with the government leads to no good.

"I don't understand why people who are involved with alternative agriculture would go to the government and ask them to legislate the standard of organics, because then it will become a bureaucratized definition and it'll probably be enforced by someone who has no connection to agriculture or organics. I see problems there. The whole certification of whether something is organic or biodynamic, I think should be dealt with by the private sector by organizations with their own standards. Besides, one organic standard through all the regions of the United

States doesn't make sense. What you'll have is an inflexible set of rules in the name of consumer safety and legitimacy. But that's just a way of taking the life out of it."

The Farm Revisited

In 1990, with the technical assistance of the American Farmland Trust, the Hawthorne Valley Farm followed through and donated the development rights to their land to the Columbia Land Conservancy. That move ensured that the land would never succumb to development pressures, and forever remain open and available for farming.

Christoph Meir departed from the farm in 1994, and moved to the Dominican Republic with his family to raise organic bananas for the export market. As the 1994 growing season got underway, Hawthorne Valley established a CSA. Steffen Schneider took on the role of farm manager, his wife Rachel took on the role of assistant gardener and CSA coordinator, and Tom Meyers assumed the responsibilities of gardener.

"The farm is working actively with the CSA approach today," Rachel Schneider explained in the autumn of 1996. "We all feel strongly that agriculture is in a unique position in the world, and that it has the possibility of demonstrating new ways of living upon the land that do not exploit or destroy resources, but rather nurture the land. Out of this can come a kind of economics that is also non-exploitive. Farming, and CSA, can help tremendously in both rural and urban areas by reintroducing people to the land. People have in fact lost touch with the natural world, and CSA can be a bridge back to reestablish that healthy relationship."

Rachel and her husband Steffen met in 1980, when Steffen was an apprentice at the Hawthorne Valley Farm, and she was the third grade teacher at the valley's Waldorf School. Tom Meyers was also an apprentice at the farm in those days.

Many of the children in Rachel's third-grade class, especially the boys, were 'city kids.' As she took them on to the farm as part of the class, she observed a dramatic change. "Something important happened," she explained. "I could see that the farm met them in a place where they needed to be touched. It came through the physical work, the responsibility of chores that they took on, and the contact with the farm animals, which the kids just loved. As a result, and I could see it so clearly, they were healthier and happier. Something very important happened, and that made a deep impression on me."

In the years that followed, Rachel and Steffen left Hawthorne Valley to work elsewhere. But they returned in 1989, as Steffen took on the job of herdsman, and then eventually farm manager. By the growing season of 1996, Steffen, Tom and Rachel, and the farm workers and apprentices, were cultivating a ten-acre garden within the farm to supply about 200 CSA shareholders, who are divided into four groups: one local, one in Manhattan, one in the Bronx, and one in Spring Valley, NY. The farm delivers the weekly shares to the remote locations.

"The shift to CSA has been wonderful for the farm," Rachel explains. "The farmers no longer feel isolated, nor do they wonder about who they are growing food for. We still supply produce for the farm store, go to the green market at Union Square in Manhattan twice a week, and sell at farmer's markets. They are all forms of direct marketing, where we know the people who are receiving the food—we work to know their needs, and they attempt to understand ours. That is important.

"A farm today, to be sustainable, needs also to do some kind of on-site processing, to add value to the foods it is growing," Rachel explains. "The Hawthorne Valley farm still has a large herd of cows. We process all of our milk for yogurt, cheese, and quark. Our bakery has grown and is very successful. The farm, dairy, green market, and bakery compose one business enterprise. We manage it collaboratively. The farm is at the center of this enterprise and is surrounded by departments that help to sustain it."

Head gardener Tom Meyers is also enthusiastic about the farm's shift to CSA. "We doubled our CSA membership over the last year, and now have a total amounting to 110 full shares, though some of them are distributed as half shares. All together we feed about 185 families. The increase in CSA membership, if anything, has confirmed that direction for this farm. We've been pleasantly surprised how open people have been to it, especially our shareholders in the Bronx. It's been very encouraging."

Steffen, Rachel and Tom collaborate in managing the farm. "We are a management team," Steffen says. "We have tried to get away from a hierarchy. We each have specific areas where we take responsibility for decisions and all, but we work together. Farming is not an easy business, and it is nice to have other people to bounce your ideas off—to get their input. It's not always easy to work together, but I believe that farmers have to learn how to do it. The day of the family farm is past. We are in a new era and things are changing. We have to learn how to work together in these new ways."

The Hawthorne Valley farmers have also benefited from a relationship of associative economy with other nearby farms. They have shared expertise and the use of heavy equipment—something that was once common in farming communities, but that has been waning in recent years as family farms have declined and industrial agriculture has come on strong. By way of example of this sharing, in 1996 Hawthorne Valley's carrot crop did poorly, while their winter squash was abundant. They learned that the nearby Roxbury Farm CSA had a good year for carrots, but was short on winter squash. They soon agreed to an exchange so that the members of both CSAs had an adequate supply of both carrots and squash.

"This is something I see coming," Tom Meyers observes. "We get lots of requests for things that we do not, and probably never will grow—citrus fruits, and nuts, and so forth. So it seems inevitable that CSAs will more and more get together and exchange. One farm can't do it all."

Some local farms near Hawthorne Valley have also started gathering together at the end of the season for a 'grower's day,' where the farmers can talk among themselves and share what they have learned from the experiences and techniques of the season. "We also study and socialize together," Rachel adds, "and that is important—to have a community among the local farmers." This kind of mutual professional support was also once common, but has waned as family farms have waned.

In an innovative arrangement, several area farms are sharing responsibility for educating apprentices—young people who come to work for room, board, and low wages to learn the many complex facets of the craft of farming. This is done under what is called the CRAFT program (Collaborative Regional Alliance for Farmer Training). "We had thirty apprentices in the 1996 growing season, in a cooperative arrangement among several farms," Rachel explains. Every other Saturday the apprentices would gather together on one or another farm for a tour of the fields and deeper education. The farmers of that land would explain what they were doing at that point in the season, and how and why they were doing it. "This took the apprentices beyond on-the-job learning," Rachel says, "and into the realm of cooperative education—where they could benefit from the knowledge and experience of a large group."

Steffen says the Hawthorne Valley Farm gets almost one inquiry a day from young people who are interested in the apprentice program. They feel the vocational call to work upon the land, and grow food for people, but they don't see any easy way to become a farmer. The high expense of land and equipment has impeded a whole generation of people called to farming. The average age of farmers in 1996 is about fifty-eight, and according to *The New York Times,* there are not many young farmers coming behind these older ones. The CRAFT program, however, stands as a beacon for many young people—a way to master the many skills a farmer needs, and to begin to understand something of CSA as well. "CSA is a less intimidating way to enter farming," Steffen says. He believes the CRAFT program can and should be mirrored by other groups of farms around the country.

At the end of the growing season all the farmers and apprentices gather together to look back over the season, to reflect and evaluate. "One interesting thing that happened at this year's gathering," Rachel explained, "was that the apprentices wanted to know more about the farmers. So each told a bit of his or her biography—what had drawn them into farming, and kept them in it. This was enormously important and helpful to the young apprentices, who are now making important decisions and setting the course of their own lives."

As at most, if not all other CSAs, one long-term issue that remains unresolved at Hawthorne Valley is providing for the old age and retirement of the farmers. Typically in America, farmers who were ready for retirement have been able to sell their farms, and use the proceeds of the sale to support themselves in old age. But the economics of this have changed for all farmers. Since the farmers at Hawthorne Valley, and at most other CSAs, do not own or have a financial stake in the land, they will not have the land to sell when they are ready to retire. "We are still engaged with this question," Rachel explains, "and it is an important one. Most of us feel that ideally the community, rather than the farmers, should own the farm. But if the farmers are working for the community, then the community needs to find a way to work for the farmers."

Gary Lamb–The Future of CSA and Associative Economy

Gary Lamb has maintained his active involvement with the Hawthorne Valley Farm, and was part of the transition to—and the ongoing development of—its CSA component. As the editor of a journal named the *Threefold Review* (see *Appendix F—Resources*), he has published a number of insightful articles on associative economy. Gary has also kept an eye on the development of the CSA movement in general.

"In the picture I have of the situation," Gary observes "CSA started out as kind of a primitive form of agricultural economy within the larger economy. But CSA has grown and matured over the years. In its early stages the typical situation was for a number of families to circle around a specific farm. As that dynamic developed, farmers found certain needs and situations whereby they started collaborating with other farms.

"So, CSA is developing from a single farm—the main farm that a community wraps itself around—to linkages with other supporting farms. In this respect it has progressed quite a bit. By collaborating together on the producer level it is possible to create a greater diversity of products for the member families.

"CSA is slowly growing in cooperation among producers and, among other things, the sharing of risk. In the single CSA farm you might say that the community will support the farm if there is a crop failure; but if indeed there is a crop failure then the community still needs food and has to get it from somewhere. Where they have to get it is the mainstream profit-driven market. So, in that sense members of the CSA movement are still dependent on the market economy—which they are trying to get away from.

"So an important question is, how does the CSA system evolve so that it can be self-sustaining as a movement, and so you don't have to go back to the system that you are trying to avoid? Perhaps the CSA movement is just getting to the critical mass where a question like that can even be entertained.

"To me another crucial step in the maturation of CSA is the possibility of consumer groups starting to relate to each other. Just now you see little examples, little seeds here and there. Various consumer groups are beginning to wonder, what are we going to do in the wintertime? And little experiments are springing up between farms in the Northeast, and farms in warm weather climates like California, to provide winter shipments for CSAs in the Northeast. There's a little collaboration there. So for the first time, winter shares are being done with farms in California.

"Likewise, there are coffee and banana plantations and so forth that grow their crops in a good way, without chemicals, and that would be willing to try and hook up with consumer groups to meet those needs. But CSA needs a critical mass for that. Within two or three years we may be at that level, where one could economically justify having a relationship with some coffee or tropical fruit plantations in, say, Central America, where you would have enough consumers together to warrant shipping the foods in that way. We are getting close.

"So we are looking at collaboration among local growers, and we are looking at collaboration to provide off-season food. How can all of this be done in a socially responsible way? Consumers and producers need to collaborate in a conscious way. Each side needs to know where the other side is at. There is a limit to the size and scope of producer-consumer associations. But I believe it is possible to create a situation where you are obtaining all of your food in a socially responsible way using the local CSA farm as the starting point."

In January of 1994, Gary gave a seminal presentation at the annual CSA conference hosted by the Biodynamic Association. He later edited his remarks into an article entitled "Community Supported Agriculture: Can it Become the Basis for a New Associative Economy?", which appeared in Issue Number 11 of the *Threefold Review,* 1994. The article has been important for many people involved with CSA and other community-based economic undertakings. Because the article shed so much light on the developing community economy of farms, several key paragraphs are excerpted below.

"Most people, being so deeply immersed in the mainstream capitalistic economy, can only see farming—or any related activity—from a market perspective. They are basing everything on market forces, supply and demand, the invisible hand, and all that. They don't have the concept, the basic concept of associative economy: of collaborating together in a conscious way to find out what should be produced, and how much—as opposed to what can you make money producing. CSA can become the basis for a new associative economy.

". . . true agriculture, which has not been converted to agribusiness, cannot survive in an economy in which there are fluctuating prices, little consumer loyalty, escalating land prices, and the exploitation of natural and human resources. But the reason for rejecting those conditions should not be merely because agriculture cannot survive under them, but because they are inherently inappropriate and harmful, no matter what is being produced and consumed . . . therefore, it makes sense that farmers would be among the first to reject some of the main components of our present economy. But rather than trying to withdraw farming from, or insulate it against the negative aspects of our present economy, it would benefit everyone in the long run if agriculture could be seen as the place to build up an economy on a new basis.

"The most basic necessity of earthly life, food, can provide the starting point for moving from our present government-guided, production-driven market economy, which is based on competition, to an independent associative economy based on consumer needs and conscious, rational decisions between the producers and consumers.

"Our current economy is production-driven—that is, the focus is on keeping businesses alive and profitable for as long as possible even if a product is not really needed. . . . One of the dynamics that is essential to most if not all CSA projects is a dialogue between farmers and consumers regarding what should be grown. . . . In this way the farmer or gardener learns the needs of the community before planning. This is a step toward overcoming the element of chance or uncertainty in the existing market.

"In this dynamic (CSA) there arises the possibility for the farmer to be relieved of the necessity of trying to create enough profit to make a living, and to focus on offering a service on behalf of the community. It is usually the case that a farmer, or any producer, in the existing economy, no matter how altruistic, must be motivated by self-interest in order to secure a living. In the CSA movement the potential exists for the farmer to be freed from such self-interest, freed to really take an interest in the community's needs, and freed to offer to meet those needs.

"The desire to found new social and economic forms lies at the heart of the CSA movement. Its potential for growth is only limited by the participants' thoughts, feelings, and will. . . . The development of CSA to this level," Gary Lamb concluded in his article, "is not a complicated task, but it does require that people who participate be personally willing to give up old habits of thought and action regarding the production, distribution, and consumption of food. To give up old habits is not easy, but it is essential in order to create new social forms befitting the human spirit."

Example 6

Caretaker Farm
Sam Smith

Caretaker Farm is a thirty-five-acre diversified organic farm following biodynamic farming practices. It is also a community supported farm or CSA with a membership of 130 households and two permanent farmers, Elizabeth and myself, who care for the farm along with the valuable support of four seasonal apprentices. Over sixty varieties of vegetables, herbs, flowers and soft fruits are grown. A young orchard will eventually provide apples for the community. The farm also has an apiary, a small flock of sheep, a few cows and pigs, and a bakery.

When my wife, Elizabeth, and I began Caretaker Farm in 1969 on the site of a defunct dairy, our initial vision for the farm was that of a small-scale, subsistence farm devoted primarily to the well-being of our family which, at the time, included three young children. Although we considered the care of the farm as important work, we were also reconciled to the idea that the farm would never be able to fully meet our needs and that one of us would have to maintain off-farm work in order to provide additional cash income. And so I continued to teach for a number of years at a nearby community college.

But soon after we moved onto the farm and started to restore its physical infrastructure, our vision for the farm changed in two ways. We began to see it as the place that might indeed be able to meet our needs for "bread labor" as well as a place that could teach us about our primary vocation to serve and protect the earth. This led us in the early 1970s to undertake, on a part-time basis, the marketing of vegetables to local restaurants. In a few years, the possibilities of making Caretaker Farm over into a small truck farm appeared to have a real potential and so in the spring of 1975 we made the fateful decision to commit all our time and energy to the farm. It was the right decision and the farm has prospered from that time on.

Now, a quarter of a century later, we look back on our relationship

This new case history was prepared by Sam and Elizabeth Smith, who farm Caretaker Farm in Williamstown, MA.   **175** with the farm with joy and pride. We feel it has been and continues to be, along with some of the older, continuous organic farms in the Northeast, a pioneer model for the renewal of America's small farms and, through its apprentice alumni, the formative training ground for many younger organic farmers now in place.

Caretaker Farm became a CSA in somewhat unique circumstances in that prior to this radical change, it was already a widely-recognized, successful organic truck farm with a popular farmstand, a bakery, long-standing connections with local restaurants, and a cooperative grocery. So why in 1990 did we let go of that to become a CSA farm?

One part of the answer to that question is that Elizabeth and I are increasingly aware that Caretaker Farm's future hangs on a single thread —our continued and uninterrupted good health. And yet we feel that our years of work have transformed the farm into a valuable local resource—particularly in a social and ecological sense—that should not be lost simply because of health, accident, old age, or the economic forces that are destroying nature and community here and everywhere.

Another reason for our conversion to CSA arises out of the vision we share with so many people for restoring the Earth's vitality and the soul of our communities. This redemptive process has to include the transformation of the most basic of all human activities: the food system which, as it is presently constituted, is not only inimical to community but also responsible for destroying on both a local and global basis the natural ecosystems on which human life and the rest of nature is dependent. Indeed CSAs demonstrate the possibility of something which humble people all over the world yearn for—that is the redirection of the economy toward community, the environment, and a sustainable future.

Responsibility for the land and the existence of a healthy local agriculture requires the understanding and participation of many people. The world's remnant of farmers can't possibly do it alone. This is the purpose and meaning of the expression "community supported farms and farm supported communities."

Transition to a CSA, and Evolving Organization

Behind the simple story of the CSA movement, is a profound challenge and question: Is it a foregone conclusion that in the present age nearly every community—be it a village, town, or city—must abdicate primary responsibility and power in its civic and commercial life to the organizational efficiencies and dictates of a global economy and large corporations? Are there more just, ecologically sound, and satisfying ways of meeting human needs based on trust, cooperation and associations at the local level?

Initially, Caretaker Farm CSA was "farmer-conceived" as opposed to "community-conceived." But the gestation and birth of the CSA could not have occurred without our farm's founding core group known as the steering committee. The contributions of the steering committee were and remain critical to the vitality of Caretaker Farm CSA and its spiritual and educational impact on the wider community. These contributions include: the initial affirmation of the idea of the farm as a CSA, the recruitment of members, help in designing an organizational structure, and the continuing concern for the well-being of the farmers and apprentices.

As the CSA began its sixth year (1996), work was underway to ask and enable the members of the community to assume major responsibility for the operation of the farm as a community supported farm. According to the ideal organizational structure toward which the farm is moving, the farmers will be responsible for crop production, the training of apprentices, and the farm ecosystem. The members of the farm will be responsible for the farm's membership, economic viability, and continuing evolution as a vital civic association within the wider community. Accordingly, the Steering Committee has initiated a process of transforming itself into a more permanent body or "Board" to provide oversight and general direction for four central working groups as follows:

- Administrative Group: Responsibilities include the preparation of the annual membership proposal; membership recruitment; and facilitation of steering committee meetings including agendas and minutes.
- Community Support Group: Responsibilities include communications with the membership (e.g. newsletter, handbook); organizing members to assist in weekly food distribution and on-farm help as needed; and fund raising activities.
- Future/Long Range Planning Group: Responsible for "mission statement"; associative relationships and community outreach; on-going renewal of the vision of the membership of its vision for the farm and the farm community.
- Farm Economy Group: Creates the annual budget and monitors income and expenses.

An Outline of Production, Membership, and Distribution

- Carrying Capacity: The total area of the farm is thirty-five acres of which seven acres are tillable. The other twenty-eight acres comprising pasture, woods, and wetlands is not suitable for tillage. In order to maintain a sustainable level of production, the tillable land is under a rotation which divides it fifity-fifty between vegetable crops and green manure/cover crops. Thus, the "carrying capacity" of the farm is three and a half acres of cropland per season.

- Membership: Consists of 240 adults plus approximately 120 children. The membership limit of 240 is determined by the carrying capacity of the farm and the operating principle that the members regard the farm as the primary source of the vegetable component in their diet. Except for special circumstances, all adults are full members of the farm and, regardless of what they take from the farm, share equally in the support of the budget. The request that all share equally in the budget of the farm sometimes occasions member dissatisfaction. Like the parable of the generous employer (*Matthew 20:1-16*), some members feel that altruistic principles should not take precedence over principles of equity and natural self-interest. But, as explained below, the farm tries to address and satisfy problems that come with trying to balance altruism with self-interest (see "Membership Concerns" below).

- Projected Crop Production: The average annual crop yield, not counting flowers and herbs, is 95,000 lbs. (or about 27,000 lbs/acre from three and a half acres) of primary vegetable crops and soft fruits. This works out to 395 lbs. per adult compared, according to USDA data, to average per capita consumption of 266 lbs. per year. Though quantities of individual crops vary from season to season, members have generally found that the farm easily supplies all the produce that their families can eat during the growing season plus a plentiful surplus for home freezing and preserving supplemented by the long-term storage crops distributed from the farm's root cellar from November to March.

- Budget and Level of Membership Support: The budget, which is prepared by the farmers and core group with subsequent ratification by the full membership, is divided equally among the 240 adult members. For 1996–97, each adult paid $310 towards the farm's budget of $72,657. Caretaker Farm CSA is not supported by shares or fractional shares. Rather the farm offers "household memberships" with the cost based on multiples of the number of adults in the household (e.g. at $310 per adult, a household membership of two adults contributes $620).

- Membership Concerns: If there is one common, perennial complaint, it is that "there is too much food." This problem occurs, in part, because crop production targets per adult are set at a level somewhat above the average level of consumption per adult.

"Surplus" production is intentional for a number of reasons. First, keeping in mind the vagaries of nature, it is imperative to till for a small "surplus" in order to assure an adequate, year-by-year level of production. Second, crop production within the CSA needs to be planned according to the natural diversity of household needs (e.g. children or varying rates of metabolism) rather than solely on the quid-pro-quo basis that operates in the marketplace. Third, assuming that members will take only what they need and leave the rest, that there may be a small weekly surplus available for donation to the local food bank or hunger program.

There is a basic misunderstanding in the comment "There's too much food." There isn't too much food just as there isn't too much healthy soil, forestland, biodiversity, or clean water and air. Indeed, there is no real "surplus" coming out of either the farm or nature. There is just enough. And so when members come to the farm every week and see what appears to be too much food, hopefully they will remember not only that, regardless of the level of the household's needs, CSA membership is a "good-buy" relative to supermarkets, but also the principle of Caretaker Farm CSA embodied in the words: "You are invited to partake in the abundance of Caretaker Farm but you needn't feel obligated to take more than you need."

In spirit, members are not "purchasing" a fixed amount of food. Rather you are contributing to the well-being of the farm community of which the soil, plants, animals (domestic and wild) and people are all an integral part. In other words, the membership contribution is not in exchange for food. Each adult's contribution is for keeping the land and restoring community. If we, the whole community, do that well, then food, which is the gift of the land, will always be nourishing and adequate for all.

To paraphrase Wendell Berry, CSA is nothing if it does not liberate farmers to be the best possible farmers they are able to be. The liberation of farmers is of paramount importance because agriculture, of all human activities, has had far and away the most significant impact on the quality of the earth's soil, air, and water.

What does good farming require? Although a small, mixed, northern temperate zone farm with a primary emphasis on vegetable crops, Caretaker Farm nonetheless follows some basic universal principles:

- Keeping tillage to a minimum in both time and space. This is critical not only to prevent erosion and protect the soil ecosystem but also to preserve the natural cover which is the basis for maintaining species diversity on the farm. While vegetable crops unavoidably require some tillage at certain stages in their cultivation, the farmer strives to keep the area of tilled land to the absolute minimum. In terms of space, one is working towards sustained, high levels of biological activity within the soil so as to support the greatest possible crop plant population on a given area. In terms of time, the farmer aims to be so precise and well-organized in scheduling of tillage, planting, succession planting, and cover cropping that when one crop comes out another goes in.
- Self-sufficiency in fertility. Although the farm is a drop point for some off-farm inputs for composting such as leaves and food wastes from the local community, the farm is moving towards as closed a system as possible through the planting of legumes, practicing crop rotations, and keeping animals for their special role in nutrient recycling.
- Internal biological control (IBC). IBC means that the control of diseases and pests depends on the health and vitality of the ecosystem which in turn depends on the level of species diversity and their complementary relationships within the farm. Species diversity is not enough. The farmer has to think also about taking advantage of the natural integrities that exist in all their diversity. In a polyculture of plants, one species that fixes nitrogen complements others that are unable to fix nitrogen. Another species may do a better job of pulling up trace elements necessary for the nitrogen fixer and other species as well. This is complementary diversity. The two are equally important, but without diversity we lack the basis for the complementary work.

These universal principles are, perhaps, the only basis for a secure food system—a system that can face anything and still carry on.

Example 7

Roxbury Farm CSA

Based upon the Roxbury Farm CSA Handbook for the 1996–97 season, prepared by the farm community, and also on interviews with Sarah Milstein and Jean-Paul Courtens.

Roxbury Farm in Claverack, NY is one of the most successful CSAs in the United States, with more than 600 shares and six, solid years of successful operation. The farm's supporting CSA community resides not just in Columbia County, where the farm is located, but also in two major metropolitan areas: Albany and New York City. In addition to its pioneering role in bringing the CSA concept to cities, the Roxbury Farm has led the way with several important CSA innovations, and established a possible model for the "family farm of tomorrow."

Interestingly, the pathway taken by Roxbury Farm's food, from the fields to the shareholders' pantries, is one of the oldest corridors of commerce in North America. The Hudson River Valley runs North and South, from the capital city of Albany to New York City and the wide Atlantic beyond. This corridor has been used for hundreds, if not thousands of years, to move food and other trade goods. Before the Dutch, English, and other European settlers, the valley was used by the Haudenausenee (Iroquois Six Nations), the Shinnecock, and other Native American Nations. Now swift roads and railways follow the ancient pathways that ran parallel to the river. The river itself is broad and beautiful, bearing vessels that range from canoes, to barges and ships.

Roxbury Farm CSA is just making another use of the Hudson River Valley, a modern way to supply sustenance to the cities, and bring necessary support out to the countryside and on to the commonwealth of their community farm. "It's been a very obvious thing for me," farmer Jean-Paul Courtens observes. "New York City has a tremendous impact on who we are, as does the Capital District of Albany. On its own

181

Columbia County is a very poor county. We don't have a lot of industry here. Much of what is here, especially the high-tech, is oriented to the global market. The farmers who are somehow surviving depend on going to the Green Market in New York City. But the dairy farmers here are tied into the global market, and they are hurting. It seems that those who are regionally based have more stable markets than those who are tied into the global market. You can see the evidence all around you that the urban-rural connection is vital. I have no qualms about thinking of CSA and the city. It is a new endeavor, but the reality of it is, there is a very strong sense of community in the city. That's a resource for farms."

Sarah Milstein agrees. She is a member of one of the three core groups for Roxbury Farm, and an active CSA networker on the national scene. She believes that "a real opportunity for CSA lies with city membership. People have said to me that within twenty-five years it's entirely possible that farming in the Hudson River Valley will be extinct—that the pressure will be too great from taxes, development, the global market, and so forth. But I think CSA presents a real opportunity for the future for farming in the Northeast, and in the Hudson River Valley particularly, where the future of CSA relies on the relationship between urban and rural communities. I'm increasingly convinced that while many CSAs will be able to exist and find support in their own areas, the real opportunity for CSA lies in their relationship with people and communities in the city. So, one can see this Hudson River Corridor as presenting a rich future, as well as a rich history. It's critical that CSA farms work together in this, rather than in competition, and there is vast potential for that to happen."

By way of example, Roxbury Farm has prospered by linking the farmland of the country with the consumer base of the city. As expressed in the farm's 1996 member handbook, "by establishing a direct relationship between the growers and consumers, the intermediary costs of transportation, processing, storage, packaging and marketing, which account for approximately seventy-five percent of the average food price, become negligible. Since there are no middle people, shareholders pay the true cost of their food, and the farmers receive the full consumer dollar."

A Trusting Relationship

The land along the wide and steady Claverack Creek has been a family farm since the 1800s. It was purchased by James and Helen Cashen in 1951, and in the early 1970s was passed on to their sons Jim and Tony.

In 1986 Jim's daughter Kelly met Jean-Paul Courtens, a young Dutch farmer. They eventually moved to the area, working first at the Hawthorne Valley Farm in a nearby town (see *Example 5*). By 1989, Jean-Paul and Kelly were ready to begin their own vegetable farm, and were offered a five-acre field at the Roxbury Farm, while the rest of the land continued to be farmed by neighbors.

Today Roxbury Farm is tended by an interrelated group of talented and dedicated farmers. Jean-Paul Courtens is the head grower, and Betsy Cashen, his sister-in-law, is the assistant gardener. They work in partnership with Jean-Paul's wife Kelly, who is raising three children. They are assisted by many people, including part-time workers and apprentices during growing season.

Jean-Paul and Kelly's first season in 1990 was tough, but proved good enough to pay back the loan that Jean-Paul and Kelly had taken out for farm equipment and supplies. Produce was sold to distributors in the Hudson Valley and in New York City, and through a local farmstand. That winter (1990–91) Jean-Paul was invited to a meeting of the NYC branch of the Anthroposophical Society to explore the possibility of a Roxbury Farm CSA project with supporting farm members from the city. He was initially skeptical of the concept, but was intrigued by the group's interest in forming a trusting relationship with the farmer, their willingness to explore possibilities, and their commitment to their ideals. So the Roxbury Farm CSA began.

Then in 1991 the Committee for Peace and Justice of the Albany Diocese of the Catholic Church came to visit the farm. The following winter, Kelly's dad, Jim, wrote to the bishop of the diocese, suggesting that the CSA model would give the church a tool through which they could actively support a small farm. That letter led to a second visit by the Committee for Peace and Justice, which was by this time ready to take action. They decided to form a relationship with the farm, and use it as a pilot project to give a practical example of how churches could support local farming. Soon, 100 Albany area families had signed up as members of the farm, and four local drop off sites were designated in the Capital District.

By the farm's second year as a CSA it had grown to 200 households, and was also wholesaling produce to several area institutions. From there Roxbury Farm has grown to its present size of about 145 acres, 30 acres in vegetables, feeding about 600 members in three geographically distinct communities. While linked by the farm, each of the three com-

munities has its own character, and its own core group to do the work of
planning and coordination.

Over the last few years the Cashen family has been meeting for
ongoing conversations about the future of the family land, and the ways
it can give security to Roxbury Farm, while preserving the possibility for
family members to build homes on various pieces of the land. With the
help of Chuck Matthei of the Equity Trust (see *Appendix B*), and the
Columbia Land Conservancy, they came to a decision.

Jim and Meg Cashen (Kelly and Betsy's parents) decided to place a
conservation easement on approximately 160 acres of the property. The
easement will stipulate that no development, other than farm-related
buildings, can occur on the farm land. Some non-agricultural land will
remain available for building a few houses.

The farm land and barns are being gifted to Betsy, Kelly and Jean-
Paul over the course of three years to take advantage of federal tax laws
which limit the amount of a non-taxable gift that may be made to any
one individual in a given year. With this decision and action, Roxbury
Farm is moving from a sole proprietorship to a partnership.

Why The Farm Is Successful

Many members of the farm community say that a big reason the CSA is
successful is Jean-Paul Courtens: his talent, knowledge, and drive. Jean-
Paul acknowledges his contribution, but sees other aspects as well: "Most
farms are seen as a business. I see that, too. There is a business aspect to
the farm, and I understand that. I have a vision, and I can organize and
manage. The general community of the CSA membership is very important, and they take responsibility for many activities. But the community
that really keeps this farm going is my immediate family. If it wasn't for
the family, then even with all the skills that I have as a farmer and a manager, this farm would not happen. There are so many things, and even
the seemingly little things are very valuable.

"First of all, the land was in the family. I didn't have to go out and get
a mortgage to buy the land. My wife Kelly and I had no savings or
resources at all when we started farming for market. We had to go to the
bank, and the bank as much as laughed at us. It was only through the
co-signing of the bank note by the family that I was able to get a loan.

"The farm is your life. You do it twenty-four hours a day. You don't
come home and forget about your work. It is always there, in every little
detail of your life. You need a strong marriage, and a strong support system behind your marriage. For instance, through the family we have

infinite baby sitters here. We have people who can run errands to the store. There's an emotional support system that's present, through three generations living now on the farm. These may seem to be small things, but they are important. I cannot put enough emphasis on this, how much it has helped, how much it has given this farm an advantage over other farms who are without that kind of support. It is a tremendous burden on them.

"A lot of farmers are farmers in the traditional way: they farm out of the heart, which is right. But you have also to think about the farm as a business. You have to run it in a certain way. Through the plants and animals you are working with the spiritual world in many ways, but, especially when you are hiring and managing people, and dealing with a market and interacting with the government through taxes and so forth, you have to think carefully about policies, procedures, protocols. These are things that many people haven't even considered in relation to farming. But once you create the systems and get them in place, it creates a freedom. Then you just cruise. Things almost go by themselves.

"Take the simple example of a hammer. It must have a place on the farm, and be kept in its place; otherwise you are going to be looking for it for half an hour. It's really that simple, not just for tools, but for the many complex dimensions of a farm. I think that that attitude about a farm is important: seeing it as a business, not in the business of just making money, but the business of being a business. It's a very different approach. I don't care about money one bit. Of course the farm needs some and I need an income, but I don't want money to consume my life, or to consume the farm. This attitude, and attention to the business, has really made it possible for Roxbury to grow really fast in a short time.

"An important part in the development of Roxbury Farm is the professional attitude, and radiation of the cohesion on the farm. It has made it easy for the CSA communities to understand and to lock into the farm. I have a sense of confidence about who I am, about the farm, and where I want to go. I have a very strong vision about where the farm needs to go. I can share that with the members. Setting the stage with this level of confidence, and a clear vision about where the farm's going to go, makes it easy for the CSA community to come in around it.

"Our vision, really, can be boiled down to one sentence: we want to create a biodynamic farm. That's it. Our mission statement, developed with the help of our mentor Hartmut Von Jeetze, who has been very helpful to us, is a bit longer. It has three parts. First, to create a biodynamic farm, not just to sustain the land, but to revitalize it; that's the advantage of biodynamics over other kinds of farming. Second, we want to create a farm which can continue beyond us, for our children or other people; that's where the CSA part comes in, and it also supports us in having a rich social and cultural life as part of the CSA community. Third, we want a sound financial enterprise. So it's that simple. This is what the focus is, and out of that comes a whole bunch of other stuff. But we have always to come back to that: this is about creating a biodynamic farm, and all that a biodynamic farm means. So when an idea comes up, we can ask: is this moving us closer to that, or away from it?"

Making It All Work

Roxbury CSA has pioneered several organizational innovations, which have been important in contributing to the farm's smooth operation and success. Several of them are outlined below, as described in the farm's *1996 Member Handbook,* much of it written by Sarah Milstein:

Share Distribution—The farm delivers produce each year from early June through late February. They deliver at their farm for members from Columbia County, and in two other locations: New York City and the Capital District of Albany. These deliveries to communities within the cities might seem daunting for some farms, but they work well for Roxbury Farm.

During the 1996–97 season, there were approximately 650 single (or half shares) supporting the farm. This is enormous by the standards of most CSAs, and likely the upper limit for Roxbury Farm. NYC has 230 shares (about 200 households); the Capital District has 260 shares (175 households), and in their home Columbia County there are 160 shares (138 households). Each site has a member who serves as Site Coordinator—someone who is primarily responsible for organizing and overseeing the vegetable distribution at their site. Often Site Coordinators are compensated for their work with a half or a whole share (this issue is decided by each Core Group).

Volunteers from the membership sign up for different shifts setting up the site, checking people in, and cleaning up. The Site Coordinator is available during the distribution to answer questions, and to provide information about the farm. After the distribution, the surplus food is donated to a local food pantry.

Until 1996, Roxbury Farm was the only CSA delivering shares to New York City, but during that year Core Group member Sarah Milstein and a group she works with called Just Food, helped support the establishment of six other CSAs in the New York City region.

The Core Groups—In addition to the growers, a number of members from each of Roxbury Farm's three distribution areas volunteer to be on their area's Core Group. The Core Groups act as steering committees, which are responsible for almost everything that happens beyond the farm gate, freeing up the growers to concentrate on growing. The Core Groups organize the vegetable distributions, oversee weekly work shifts, enroll new members, do the administrative and treasury work of the memberships, plan the farm festivals, and hammer out the budget and share prices.

While the individual Core Groups meet on their own schedule about once a month, all three groups and the growers get together twice a year for Summit Meetings to discuss the overall direction of the farm and any shared activities, such as the farm festivals, or shared responsibilities—such as the purchase of a new cooler. The Spring Summit Meeting, held in April, generally anticipates the coming season, while the Fall Summit Meeting, held in November generally reviews the previous year. The Core Groups take turns convening and facilitating the Summits.

Budgetary Matters—The creation of Roxbury Farm's yearly CSA budget represents a collaborative effort between the farmers and the members. The farmers and representatives from each of the Core Groups meet for one full day in the late fall to go over the past year's expenses, and to project the expenses and needs for the coming year. This annual budget meeting focuses on two areas: the farm operating expenses (what it takes to grow the food, including the farmers' compensation), and funds for capital improvements and long-term stability. These two items are combined to determine the total cost of producing the food and supporting the sustainability of the farm.

Once this overall figure is determined, it is divided by the total number of member shares to arrive at a base price per share. The farm's yearly budget (in 1966) of approximately $200,000 was covered 80 percent by member pledges. The remaining 20 percent was covered by sales to several nearby institutions that serve meals regularly, to health food stores, and by the purchase of meat shares by some of the CSA members.

In addition to the farm-based budget meeting, each of the Core Groups meets separately to work out the administrative costs for their individual communities. These costs might include compensation for site coordinators, postage and photocopying, equipment for the site, and so forth. These administrative costs are different for each of the three communities, as are the vegetable transportation costs, which are determined by a formula that accounts for gas, wear and tear on the truck, labor hours, and so forth. Each community totals its administrative and transportation costs, divides by the number of members locally, and adds this to the base price per share for their community.

As an important point of education for all members of the farm, the fall re-enrollment process in each of the three communities includes an explanation of the budget, and how the share price was arrived at. In this way all members have an opportunity to understand how the price they pay for a share is a true reflection of the needs of the farm community.

Communication Tools—In June, 1993 the Roxbury Farm community in the Capital District of Albany began publishing a newsletter for their members, distributed at their four local pick-up sites. By October of 1994 the newsletter was serving all three of the communities that are linked in their support of the Roxbury Farm. The newsletter reports on farm events, pertinent experiences of the members, CSAs in general, biodynamic farming, agricultural issues, and other events and topics of related interest. The newsletter is printed and delivered to all pick-up sites each month during the summer, and every other month through the fall and winter.

In addition, each week a handwritten letter from one of the farmers accompanies every share, to relate the news of the farm that week. Roxbury Farm has found that this builds the members' understanding and appreciation of the work of the farms and the role of the natural elements in growing vegetables. The handwritten aspect of the letter gives farm members a closer feel for the people who are providing their sustenance. On the back of the weekly letter appear recipes for some of the vegetables that are part of the week's harvest. When an uncommon vegetable is in season, the recipes are much welcomed.

Harvester's Share—Often the most time-consuming part of raising a food crop is the harvest. The Roxbury Farm CSA has found this to be especially true with strawberries and sugar snap peas, which are ripe for picking June—which is also the time for the first cutting of hay, and major planting and weeding activities.

Other CSAs which have most of their membership picking up shares right at their farm have instituted pick-your-own programs to reduce the amount of time the farmers have to spend on harvesting; however, the Roxbury Farm CSA was not able to offer that option because about seventy percent of the farm's members get their shares delivered to urban sites far from the fields. In response to the labor squeeze, the Roxbury Farm CSA came up with an arrangement known as the Harvester's Share.

The Harvester's Share works for those farm members who are able to make a commitment to come to the farm any or all weekday mornings in the month of June to harvest sugar snaps and berries for the membership at large. The amount of produce picked is weighed, recorded, and totaled for the whole time period of labor the member has worked, and the farm then recognizes a monetary value of forty cents for a quart of strawberries, and fifty cents for a pound of sugar snap peas. The total dollar value is then deducted from the cost of that member's share.

The farm estimates that to complete the harvest in June it needs seventy-two hours total, per week, from members doing the Harvester's Share. Generally one of the farmers or apprentices works along with the members. They recommend that the members who come bring sun protection, such as lotions and a hat, and also a water bottle.

Shareplus—The Roxbury Farm CSA also has a program called Shareplus, through which farm members may place bulk orders for extra vegetables. The program is a supplement to regular membership, and does not change the normal membership in any way.

Shareplus has three important benefits. It allows participating members to extend their consumption of the farm's harvest through the winter by canning, freezing, pickling, or drying additional produce. It also provides the farm with a welcome flexibility in the chancy enterprise of growing and harvesting vegetables. Finally, it gives the farm an extra measure of community-based support.

During the 1996 season, the cost of participating in the Shareplus system was sixty-five dollars. The farm offered a participation agreement, listing ten crops that are Shareplus options, and detailing how much of each crop would equal one unit. Participants pick any seven units of the ten crops, and rank their choices in order. A member could, for example, choose seven units of the same crop, or distribute the choices among different crops. The farm then tries to fill the top four or five choices of every Shareplus member. In the event of a crop deficiency, the farm uses the members sixth and seventh choices to fill in the gaps. As an added bonus, participants may receive surplus vegetables not listed on the order form. In 1996, for example, excess lettuce, kale, and various herbs were available to Shareplus members.

Local Co-Op Buying —The Roxbury Farm CSA membership currently supports one family farm. But, as acknowledged in the farm's handbook, by applying the CSA concept and infrastructure elsewhere, there is potential for membership to help support other growers in the local area. The Roxbury Farm CSA has established relationships with about a dozen nearby farms, and offers the Columbia County members the opportunity to buy everything from eggs to honey. The Capital District community has a link with a Schoharie, NY dairy farm, from whom they buy yogurt, and a bread baker who is associated with another CSA.

Example 8

Good Humus Produce CSA
Annie Main

Watching the men, women, fathers, friends, uncles and grandmothers around us tilling the soil, tending the plants and harvesting the fruits of their labor, has always been a part of both my life and my husband's life. I grew up watching my uncles spraying sulfur on their wine grapes in the Sonoma Valley, and each summer I found myself on hands and knees under the prune trees, plucking those fruits off the ground and putting them into what seemed like bottomless buckets. My husband Jeff's first job as a teenager was cutting peaches for drying in the Central Valley of California, and then going home with a few extra dollars in exchange for those long hours of fuzzy, itchy peach cutting. The people we saw around us were solid people with muscles, determination, and an ability to survive the fluctuations of farming. This created an image for us of people doing work that was respectable, honest, and necessary for their communities.

Our college years brought us both to a university town in a valley that was, and still is, permeated by agriculture. Neither of us studied agriculture in school. Our living situation influenced our educational experiences through cooperative living, growing a large garden for a ten-person household, and participating in a program to educate others on food growing and preservation. This was a time of great change in the food system, and we both became involved in that change. We helped our food buying club expand into a cooperative storefront, and several of us also came together to create one of the first farmer's markets in California. We saw there was a need for small farmers to sell directly to the people who eat their food.

Without knowing what imprints on us, what gives our lives direction, my husband and I found ourselves planting our first garden together in our first year of marriage, managing a farmer's market, buying produce from the local wholesale markets for the new food coop, and caretaking the relic of a mid-1800s grain farm. What remained at this farm was antique equipment, and many stories of the Central Valley past: stories of mule teams, steam threshers, thousands of acres of wheat, the early years of agriculture in the region, and the people who lived that life. What we realized was that managing a farmers' market, volunteering our time at the coop, and caretaking the stories, was not enough to make ends meet.

In 1976, three families from the market and coop came together. We created our first organic farming business, a partnership on three-quarters of an acre. Thus began Good Humus Produce. At this time organic farmers were almost nonexistent. Wholesalers of organic produce were just getting settled into the business, and organic produce was often holey, small, or old. This was the beginning of a farm with all that needs to be learned, and also the beginning of our work in life. Indeed, we worked. We lived to farm, and farmed to live, surviving from month to month, year to year.

By our third year we had expanded from three-quarters of an acre of organic vegetables, to ten acres of vegetables, seven acres of fresh market apricots, and one acre of U-Pick boysenberries. These were our young years, the giddy times, when we pushed our bodies to the limit time and again. We spent hours philosophizing while picking summer squash on a Sunday morning, or while resting under the tall, intoxicating foliage of the fava bean cover crop.

As our farm grew, so did we. Our friends started farming, our community became larger, and we became familiar faces at the markets. Eight years later, with help from an old mentor, we found ourselves with land to plant our own orchard, and to grow our vegetables, build our home, barn, and family. We were like moles, faces to our work, noses in the soil, tending to farm and family. We were feet, body and mind upon the earth, living lives of soil and toil. With the arrival of children we paused and saw that there was more to the farming life than work. New concepts were coming forth, reinvigorating old ideals of cooperation and community. Farming friends around us were starting Community Supported Agriculture programs on their farms. While we were helping to create a farm at our childrens' Waldorf School, we invited Steve and Gloria Decater from Live Power Farm in Covelo, CA, pioneers of CSA in California, to introduce us to the concepts and practices. In 1993, the school and our farm each began a CSA program.

Living in the West, where new ideas and markets come and go in the produce business, we have found that the way for us to survive is to be a diverse farm with diverse marketing outlets—and to always look for niches. CSA began for us as an additional marketing strategy, along with farmers' markets, retail, and wholesale outlets. As small family farmers we were not willing to sacrifice all the relationships we had built up in the marketplace over the years for what seemed just a new marketing idea. We also didn't have the faith that a community would be willing to support us for an entire year. We created a subscription CSA, as did many of the farms around us.

With thirteen acres of vegetables, five acres of fruit, one acre of perennial flowers and herbs, and seventeen years of experience at farmers markets, we knew we could handle the CSA food deliveries. We love to grow food, flowers, and herbs, but in the context of our current aggressive economic marketing society, Jeff and I found ourselves beginning shyly. Our CSA membership began with our friends in town. We delivered ten pre-packed boxes of ten to twelve different items for $150 a quarter (three months). It was a small, slow beginning, with our friends afraid to hurt our feelings by telling us it wasn't working for them. We have since come to the realization that this pre-determined, prepaid weekly delivery of unknown produce is not for everyone. But what helped us to grow and stabilize were articles on the CSA concept that appeared in the food and business section of the Sacramento and San Francisco papers. The publicity helped us to increase our membership eight times in the first year. One new member was so excited when she picked up her box that she went directly to her local newspaper and demanded that an article be written.

With the California climate we are able to farm year round, and thus we can deliver boxes of produce year round, although we take one delivery week a quarter off for our own sanity. The break in delivery invariably causes confusion, so we try to break when everyone else is busy with holidays. Our CSA membership now fluctuates from sixty to eighty members, with at least a ten percent loss of members quarterly. We ask for quarterly payment, feeling that few are in the economic position to prepay for the entire year. Our largest member loss comes during the winter months as members anticipate weeks of leafy greens. We are learning a lot about our members' eating habits and abilities, which vary tremendously from ours.

The Central Valley of California is a heartland of agriculture, where food is readily available and inexpensive. Our members are thus accustomed to a large, year-round variety of inexpensive food. As small farmers with limited acreage, we look for those crops that have a high continuous production over a long season. In the winter months, lettuce, kale, collards, mustard, and other greens fit our picture of crops with a high economic return. In the meantime, we educate our members on the use of, and nutritional benefits of, seasonal produce. Education and lifestyle changes are slow, and we find it necessary to be responsive and somewhat accommodating to our membership's tastes.

The winter months, if planted right, can be a profitable harvest time. There is less competition, less labor necessary, and with the drier winter weather, an opportunity for us to plant continuously. In 1995, however, our winter changed dramatically. That year it rained for a solid month—no planting, no cultivating, a lot of rotting, and slim pickings for our CSA boxes. Interestingly, though, this wet weather created an opportunity for farm community building through produce exchange. We now exchange dried fruit, herbs, and excess crops from our farm for the excess from other farms. This gives us the ability to keep diversity, quality, and consistency in our CSA program. Each delivery is invoiced and records are kept by the farms involved in the exchange. We have exchanged produce only with our local friends, and farmers who we know, and in whom we have confidence. In our newsletter we keep our members informed about where additional produce comes from.

Subscription-style CSA has helped bring us to maturity, to our next step as farmers. As we are educating our membership, and expecting them to realize the need to connect to their food, their farmers, and to the land, we too are waking up. Writing our weekly newsletter has brought a heightened clarity of purpose to our daily experiences of living on the land, of growing the food, of living the rhythms of the seasons. Our members tell us that their bodies get fed by the produce, and their souls get fed by the newsletters and CSA membership.

I feel I have spent twenty years living the farm life, but not with a true awareness as we have concentrated on production. Our writing has helped us to articulate to ourselves the activities of the farm, of nature, of the agriculture that lives around us, and our inner spiritual growth. We view this farm not only as a place of food production, but also as a place for all of us to experience for a heightened awareness of nature and spirituality.

In our transforming years we are taking a closer look at our roots on this farm, our direction, questioning what is to come next. I feel as individuals we are looking for a place of refuge, nurturing and deepening our spiritual progress. The farm is a place to create this refuge, to work in community, to physically tend to the land, to feed our bodies and our souls. Our next step is to see how we can further mature with the life of the farm and with ourselves.

We feel it is time to ask for our community to strengthen itself. This year (1997) we will be working on bringing our CSA community closer to the ideals of community supported agriculture, and farm supported communities. We will be having more social gatherings at each designated drop-off site, along with two farm visits a year; we will be asking for continued outreach for new members; we will be asking for a longer-term commitment, and for monetary help to plant a new citrus orchard.

This year we hope also to work further with our local farming community to create more intentional cooperation. One step will be to combine all our different added-value products (such as jams, jellies, dried fruits and herbs) into a cooperative catalogue. That way all our products may reach each of our CSAs, and possibly other communities.

We plan also to increase the plant diversity by implementing a native and edible hedgerow throughout the farm. We have diverse crops on our farm, but our cropping has a two-dimensional feel to it. We would like a more three-dimensional system that includes not only fruit trees and vegetables, but also native trees that will help increase wildlife habitat. We look for shrubs, perennials and annuals teeming with life to bring more diversity to the system. We look for increased insect and wildlife populations, but also our own area of wilderness on the farm that can possibly enhance the sense of peacefulness and refuge here.

One of our longer-term goals is to reduce our dependence on purchased inputs, such as power for our irrigation water. I believe one of the most important goals we have for ourselves, our children, and those who visit or work with us, is to gain an understanding, a respect for the work, for the land, and for the care necessary to live in harmony with the world around us.

For the future Jeff and I want to share the beauty of the land—the wilderness, and the farm—and to celebrate the rhythms of the seasons with others. We are searching and asking how to bring this into our farm, and yet be able to do what is necessary to continue the maturing process of the farm and ourselves. Above all we are searching for a system that will sustain the land, the farm, and our farm family so that we can continue doing work that is respectable, honest, and necessary for our community.

Example 9

Forty Acres and Ewe
Jim Bruns & Donna Goodlaxson

When I came to CSA, I had come to see our existing public food and agricultural institutions to be unfriendly to both consumers and farmers. Without having heard of CSA, I knew my hope was in some kind of a consumer/farmer alliance. In retrospect, I think this was true for many of us; it was an idea whose time had come.

My parents had a dairy farm in Western Minnesota. Their small farm did okay financially until the 1980s. My father quit farming in 1986. It was a typical story of the '80s with debt, high interest rates, low commodity prices and some poor choices leading to financial failure. These troubling economic problems cast a dark shadow on what had been a good quality of life.

In 1991 I met Donna, my wife and partner, a "farm girl" whose family had survived the previous decade, but struggled with painful intergenerational tensions. Donna shared my vision of a consumer-farmer alliance. Later that year, after reading *Farms of Tomorrow* and visiting several eastern CSAs, we decided to accept the challenge of Verna Kragnes from Philadelphia Community Farm and "just do it!"

We felt well suited to the task with farm backgrounds and grassroots organizing experience. We had an extended urban community with similar world views from which to build membership. Donna was already living in a rural community about ninety miles from Minneapolis/St. Paul and had existing relationships with other farmers in the area. I had been doing large-scale urban gardening for several years.

Decision Time: 1992
In the winter of 1991–92, we discovered Donna was pregnant, which actually cemented our decision. We held our first organizational meeting where a Core Group was formed and a quasi-democratic process resulted in the farm name "Forty Acres and Ewe." The farm came to us.

196   Jim Bruns and his wife Donna Goodlaxson farm Forty Acres and Ewe CSA in Wisconsin.

One of Donna's neighbors was moving and wanted renters for her organic farm, which included sixty acres, an old brick house, a machine shed and a large old barn. Rent was $250 per month. How could we say no? That spring we moved ourselves and Donna's six sheep, made maple syrup, cleaned up lots of piles and plowed sod. Soon we intimately understood the term "organic by neglect."

Our vision was to have a whole farm model so we offered meat and chickens as well as vegetables our first year. Vegetable shares were large and priced at $500 each. A large "Big Meat" share was added for $200 making it the deal of a lifetime. We signed up about forty vegetable shares and several meat shares. To save on labor, deliveries were made market-style with bins of various produce. However, a killing frost on Father's Day, June 22, followed by an extremely cool summer made the season a tough one. We delivered very few ripe tomatoes, and the vegetable share value was rather low. Managing cattle, sheep, hogs and chickens along with the garden on a new farm was very difficult. We got through the season and appropriately had our first child at the end of August on a share delivery day.

Learning and Changing in 1993

Our second year we switched to a half-sized share. Most of our friends were single or couples with no children. The large share was not working well for them and they seemed to be our base. We decided to charge by the adult. If there was one adult in the household they would buy one share, two would buy two, and so on. Children would be free, all parents could take enough for their children. We also put into place a sliding fee scale that year, $125-$200 per adult. Most people opted for the very bottom of the fee scale and purchased only one share per household. Produce levels and variety were still somewhat improved and we continued with the market-style distribution. We had a wonderful Solstice celebration of our marriage after a week of ten inches of rain. We again made very little money and found that food stamps are ridiculous for organic farmers to receive.

Expanding in 1994

In year three we raised the price slightly to $135 and worked to clearly communicate the share structure. We dropped all meat except chickens so we could concentrate on the garden. This worked somewhat better and we delivered around 170 adult shares. We still grew too many vegetables for too little money. Our greatest challenge was that we were faced with losing half of our labor force mid-season since Donna was pregnant again. A friend introduced us to a family who were interested in working with us but we didn't have money to pay them. What we did have was vegetables, so we offered them thirty shares which they marketed under their own farm name. In spite of being on a steep learning curve ourselves and unprepared to really teach, we persevered. That summer we birthed another CSA, as well as our son. The fall brought the unexpected opportunity to buy the farm. We found financing through an ecumenical loan fund and did a contract for deed on some of the acreage. The result was the security of owning a twenty-acre farm with a low monthly mortgage.

At this point I began to see CSA less in its ideal form and more in its contextual form. Ideally, we should have been able to put together a solid membership base that "got it" by now. Yet even with progressive, food coop shoppers as a base, it seemed we were unsuccessful at conveying the whole concept of CSA. While the core group was strong and we had great fundraisers and events, our membership was still essentially shopping.

Our end-of-the-year surveys showed that supporting a small family farm was important, but fresh organic vegetables seemed to be the major attraction. The turnover was quite high, 50 percent each year, and it seemed the major reason for leaving was lack of economic benefit. It's interesting to note that some felt they didn't use enough of the vegetables they got, while others felt they didn't get enough. Membership seemed to have great difficulty making even the modest lifestyle changes, such as a once-a-week pick-up of vegetables. It began to be clear that the one group that was holding stable for us was families.

Growth and Expansion in 1995

With the lessons of the previous years in mind, our fourth year plans included a return to the full-sized vegetable share, this time for $350; the dropping of all meat to concentrate on vegetables; and active recruitment of families. This was the fourth time we had changed share price or size in four years. We lost several old-time members who couldn't make the change work for them. Share numbers were still lower than projected with about sixty, but the farm was more profitable and the work load more manageable. We asked members to do one day of farm work and found that once they had traveled to the farm they were much more likely to join again. Production was good and our core group was becoming quite strong, handling most of the administrative work. With the exception of low revenue, it was a good year.

Working and Needs: 1996

Farmers are known to be optimists and we were as we entered our fifth year. It seemed CSA was doing well everywhere. We had strong replies to our surveys, so we planned for a slight increase in share price to $400, and a large increase in membership from 60 to 125. Our major goal was to raise revenues through share sales. Our membership drive ended with about seventy-five shares. Consumer interest seemed to have peaked in the region. We kept a lot of our members, but we didn't get many new ones. This was common for many CSAs in the region. Produce quality and variety was better than ever. Membership was strong in terms other than sheer numbers. We implemented a "requirement" for a working visit and had 80 percent compliance. Members continued to be satisfied with what had come to be called a "need-based distribution." Suggested amounts are posted along with the bins of different vegetables. People take more of the things they love and none of the things they won't use. The core group did even more of the administrative work. Our third child was born that fall, late, after all the garden work was done.

1997: Year of Commitment

As we look to our sixth year we're moving to diversify our income base through a marketing cooperative with five other CSA farms, and by working with a few restaurants. CSA continues to be our preferred source of income. A consistent 100 share base is what we've concluded the farm currently needs to succeed financially. A marketing consultant has been hired (for a share) to help the Core Group with recruitment. Our Core Group has matured, showing strong commitment through a willingness to meet monthly and a desire to do everything they can in order to free us up to "just farm." Their transformation as a group was summed up recently by a Core Group member who said, "I finally understand enough about how and when things happen that I know what I can do to help."

Successes and Changes

From our experience we've learned to define success differently. It is certainly not measured in dollars. We have been successful at learning to be good growers and developing a sense of ourselves as part of a much larger rhythmic system. We've evolved a successful distribution system, a solid membership base, and an involved Core Group. Our family and our farming community are strong with five CSAs within ten miles. Each Friday at noon throughout the summer, we gather at alternating farms for "Farmer Lunch." We all work together as much as possible sharing ideas, milestones, strategies, disappointments, equipment, and some work. Our children are growing up together and we keep tabs on one another knowing that our success is intertwined. Most recently we formed the Hay River Produce Cooperative and are building a relationship with a chain of grocery stores. There have been times of doubt, but we've never doubted that what we are doing is right and righteous. We have a rich life.

If I had to do it all over again I'd do three things differently. First, I'd train under a market or CSA gardener for at least one season before starting. Second, I'd start with a capital base. Third, I'd keep an outside income for awhile and grow into full-time farm income.

Summary

My hope for CSA was that it would offer both consumers and farmers the chance to escape the perils and insecurities of the market by working with each other. The farmer/consumer alliance I envisioned has not materialized to the degree and at the rate I'd hoped for. I still believe this is the direction we must go. Only those who eat our food, not those who profit from it, will be our allies.

I've been most surprised by the degree to which our culture is mired in consumerism. Even the most progressive of thinkers have had a hard time changing their living and buying habits for a mere sixteen weeks a year of vegetable sourcing. For most of the people we've worked with over the years, it seems initially to be a different way to purchase vegetables, not a different way to relate to agriculture. The vast majority come and go based upon their perceived economic benefit. They come to us as consumers and it's up to us to make it possible for them to one day be our partners. This is our greatest challenge, the human, not the biological side of agriculture.

For myself, it's hard to take CSA out of the context of my life. To go from being single in the city to married with three children on a farm in five years is a lot of personal change. In many ways, the struggles and rewards of the CSA pale in comparison with those of raising three children. If, in the end, all I have to show for the past five years is that I've been with my children, would I be a poor man?

In many ways, CSA is what I'm doing while my children teach me about myself and the world. The farm is a great learning environment and I am forever indebted to them for their vision.

I'm also indebted to biodynamics and anthroposophic thinking. CSA has brought me from Wendell Berry to Rudolf Steiner. Biodynamics has the most coherent philosophy about people, nature, and farming I've yet encountered. People like Trauger Groh, Hartmut von Jeetze and Karl Koenig have contributed greatly to my present world and spiritual views.

Example 10

Variations On A Theme:
The Expanding Scope of Community Farms

From 1986, when the first true CSAs were started in America, into 1997, nearly 1,000 such farms and gardens have been established, involving perhaps 100,000 households. Each has adapted the basic CSA idea to its own realities and preferences. Some of those variations are explored in the many case examples presented in this book; yet others are briefly outlined in this chapter.

We have come to see that CSA accommodates all shapes and sizes of farms and operations—from ten-member gardens, up to full-scale farms with as many as 600 member households. Most CSA practices and procedures can be readily scaled-up or scaled-down to fit the needs of the farm, farmers, and families.

Charles Beedy, Executive Director of the Biodynamic Association, has observed that "the CSA movement is the vehicle which allows the greatest number of people to associate themselves with agriculture. Shareholders come from a wide range of ages, and races, occupations, and economic brackets. In that sense, CSA is the epitome of democracy. We are still at the beginning of the exploration and discovery of the many forms that this movement can assume in expressing itself in both economic and social terms.

"For me," Beedy says, "CSA continues to be a movement which resists single definitions. No 'how to' book can narrowly define what it is, or what it can be. It meets various peoples needs. It can work in urban situations, or even in rural situations where you don't have great population density. And yet it can still be a CSA."

While there is a general CSA theory, or ideal, as set out in preceding chapters, there is also the free spirit of the human community transcending theory or creatively adapting to harsh realities. The following sections outline some of those variations.

Congregation Supported Agriculture

Religious congregations are natural communities where CSAs can take root, since the group already shares interests, and has experience working together. Most congregations have regular weekly meetings, and most have a place, such as a church hall or basement, that could serve as a central distribution point. In 1990 Dan Guenthner of Osceola, Wisconsin attended a conference on sustainable agriculture and heard someone pose a question: "Where are the churches?" A committed Lutheran, he took the question seriously and it propelled him to explore ways that church congregations can become directly involved as stewards of the earth by supporting local farms. Guenthner has written a comprehensive booklet on the subject, *To Till It and Keep It: New Models for Congregational Involvement with the Land* (see *Appendix F—Resources*).

"Historically," Guenthner writes in the booklet, "the major religious denominations in this country have been involved in agriculture through a charitable, relief approach to global, national and local food distribution problems. These efforts have been significant in both scope and impact. As important as these relief efforts are, it is becoming increasingly apparent to many of those monitoring our current agricultural predicament that a much more comprehensive approach to the needs of the land of the people will be necessary.

"What we need is real, tangible, faith-based models that incorporate our lives and ministries with the urgent needs of the land. . . . I remain hopeful that sustainable farmers everywhere will come to see our churches not as idle spectators, but as active participants. . . . In doing so, we can transform not only our churches and our faith, but we can also come to know in a much more intimate way what it means to 'dwell in the land securely.'"

Guenthner suggests that instead of food drives, or sending a check out to a local food pantry or homeless shelter, congregations could consider a range of options:
- Converting church lawns, or other church-owned land, into congregation gardens.
- Checking to see whether congregation members have land that is suitable for a farming or gardening project.
- Starting a CSA by supporting a local farmer, who may even be a member of the congregation. Many such church-supported farms have blossomed in the US and Canada in recent years. The congregation forms a natural community, and food can even be delivered to the church hall for pickup after regular weekly services, or other

church-sponsored activities. Church buildings are natural sites for meetings of the farm's core group, and farm news can readily be incorporated in the churches bulletin or newsletter. The church setting is a natural for pot-luck suppers with the farm workers, and the farm or garden can provide a natural setting for church festivals.

• Involving the elders and the youth of the congregation in various garden or farm projects.

Some churches have been able to see tangible results, with their interest and support of a CSA resulting in the protection of a local piece of farmland, and ensuring that it is farmed ecologically. Some churches distribute the excess harvest to low-income citizens who otherwise would not have access to fresh, locally grown fruits and vegetables.

The National Catholic Rural Life Conference (*Appendix F*) has also surveyed the movement, and found that many of its congregations and religious communities around the nation have undertaken CSAs. "Deep within the ecological crisis," they write, "is a spiritual crisis, at the heart of which lies the spiritual error of excessive materialism." Their report includes a number of case studies showing a variety of creative practices in sustainable agriculture and sustainable community. As the report states, some Catholic communities are learning how to live within the limits and possibilities imposed by their place and circumstances.

"Several religious congregations have established some kind of land-based center for purposes of education, demonstration, and outreach. . . .They clearly have a public role beyond that of change agent within the religious order, for they also act as a moral force for just and sustainable living in their regions."

However, some CSAs have had poor luck in their efforts to establish relationships with religious groups. Much apparently rests on the enthusiasm of the religious community leaders for the concept. If one group is not responsive, another may well be. Based on cumulative experience, groups interested in this approach may have to knock on several church, temple, or synagogue doors to find a congregation that understands the concept, the importance, and the possibilities.

Corporation Supported Agriculture

Some farms have forged links with businesses to provide a supply of fresh food for employees, and to bring agriculture closer to their workers. In many respects, the workplace is a key modern community center. Instead of a traditional community of neighbors, people often form the closest community bonds with the people with whom they share desk or workshop space.

For example, Patagonia, Inc., a company which manufactures and sells outdoor clothing and sports equipment, has developed an innovative relationship with Fairfield Gardens near Santa Barbara, California.

Fairfield Gardens is located in the midst of a densely developed area. The growing fields are in fact surrounded on all sides by a suburban tract development. Adjusting to this reality, master gardener Michael Abelman founded the Center for Urban Agriculture several years ago to promote agriculture in urban areas, and then Fairfield Gardens also became a CSA.

Jill Villigen is Director of Patagonia's Environmental Programs. She says that in 1995 the company helped to formally coordinate the participation of employees in the Fairfield Gardens CSA. All together, the company was receiving twenty shares a week. Some employees purchased whole shares, and others teamed up to split shares. Since the corporate headquarters is located about thirty-two miles from the garden, the CSA delivered the shares to the company. "From the company's perspective," Villigen says, "the program worked well. There was a lot of interest, and a lot of satisfaction. It expanded the understanding of our employees about the linkage between the farm and their dinner tables."

Gardener Michael Abelman, however, was somewhat disappointed in the way it worked. One difficulty was that many of the corporate employees involved with the CSA had to travel frequently on business, and thus could not use all of the food that came with a share. "In some cases, a single share was being divided among five households, and the shares were just not designed to be broken up that way. It was tough to please everyone.

"We haven't given up on it yet," he said. "Many of the company's employees are involved as CSA members this year (1996), but they pick up their shares at the garden and participate as individuals rather than being coordinated through the company. We will see what happens in the years ahead."

Beyond participation in the CSA, Patagonia has supported the farm in other ways. In 1995 Fairfield Gardens benefited from Patagonia, Inc.'s Environmental Program grants program. Every year the company distributes one percent of its profits among 200 to 300 different grassroots environmental programs. That sum is usually one to one-and-a-half million dollars a year. Fairfield Gardens was one of the many beneficiary groups in 1995, and may benefit again. This is support the garden is much in need of, for it is actively seeking to raise about $750,000 to purchase the land on which it is located.

Also in 1995, Patagonia held its annual all-employee meeting at Fairfield Gardens. During the day, employees pitched in with various projects, such as helping to restore the creek that runs through the garden by clearing alien plants that had invaded the waters, and were threatening to choke the flow and crowd out native species. Over thirty company employees also volunteered to help the Garden host an open house that gave the public an opportunity to learn about urban farming.

College or School Supported Agriculture

Colleges and universities have also been studying and establishing CSAs. For example, The Dartmouth Organic Farm is located at the Dartmouth College-owned Fullington Farm, in Hanover, NH, three miles from the main campus.

The idea for a school-supported CSA originated in an Environmental Studies class in 1991, and was slowly developed until the first year of full operation in 1996. The farm employs a professional manager who not only oversees farm operations, but also coordinates student research, promotes academic use, and stimulates volunteer efforts. The farm provides organic produce for Dartmouth's dining services, for the Hanover Inn, and also for the school's Moosilauke Ravine Lodge, as well as for students and employees through a garden stand set up at the college's Collis student center.

The plan is for the college to support the farm for the first three years (1996–99) with funding of $30,000, then $18,000 and finally $8,000 in the third year. The budget was set with the assumption the farm would be self-supporting by the fourth year through revenues from shares, produce sales, and outside donations.

As a part of Dartmouth College, the farm is designed to further the college's educational mission in four broad ways:

- By providing an interdisciplinary educational experience which encourages a wide range of students to learn and to work together in the outdoors.
- By providing a variety of thesis and culminating-experience projects, which are now required of all students as part of the college's new curriculum.
- By fostering student interest in ecology and natural history through the development of practical gardening skills.

- By emphasizing the college's long-standing commitment to environmental education and student interest in environmental issues.

The farm, administered through Dartmouth's Outdoor Programs Office, is advised by an ad hoc committee of administrators and faculty from the college's departments of Biology, Earth Sciences, Environmental Studies, and Geography. They work with the farm manager to set yearly objectives, and to integrate the farm with the college's overall curriculum.

One unexpected benefit of the project is that the farm has created an opportunity for Dartmouth's many international students to grow crops native to their homelands, which otherwise would be unavailable. The farm also hosts programs from the greater Hanover community and the Upper Valley region of New Hampshire such as composting, greenhouse design, and gardening.

Many students use the farm for other educational projects, such as biological research, environmental observation, a solar-irrigation project, and soil mapping to fulfill the requirement of a geography course. Creative projects also use the farm as a base: one student prepared an oral history of farming in the area, another obtained support to write poems based on the life of the farm, and yet another researched traditional agricultural practices of the indigenous Abenaki and Iroquois nations.

Community Supported Composting

Many CSAs are exploring Community Supported Composting (CSC) programs, and have found the programs to be a valuable source of soil-building material for the garden or farm, as well as a responsible way for communities to recycle food scraps, leaves, and lawn clippings from the town or city, thereby easing pressure on local landfills. Composting can build bridges between urban and suburban residents, who provide the raw materials for compost, and the farms and gardens which put that compost to work in the soil.

"Implicit in the CSA movement is the building of the soil as a vitally important component," observes William F. Brinton of Woods End Agricultural Institute in Mt. Vernon, Maine. "Intensive vegetable culture takes a lot out of the soil and farmers need to put it back." Through CSC, the community can provide the necessary organic material, and also take pressure off local landfills.

Mr. Brinton has systematically explored the reasons for, and the possibilities of community supported composting. Over the last several years, Woods End has worked with many community outreach programs centered on the solid waste crisis. They have found that many consumers care deeply about becoming involved in the cycle of nature, and in returning organic materials to the soil.

The Woods End Institute approaches communities, schools, colleges, camps, and other institutions to develop their own recycling and composting systems, and then helps tie them in to organic or biodynamic farms. They have found that in schools the students are usually eager to become involved, especially once they have tracked the food chain through their cafeteria to see where it comes from, and then where the waste goes. Through a CSC they can make a decision to direct the food waste stream back directly to the earth at a local farm, a farm they may even visit on a class field trip.

When the solid waste crisis hit in the 1980s, interest in composting skyrocketed. However, it took off in a primarily economic direction, Mr. Brinton has observed. The composting movement was driven primarily by the need either to save or to make money, and the innumerable permutations of that basic equation. Much of the same thinking that lies behind the commercialization and industrialization of everything—including farming—was also underlying the resource recovery and composting ventures that came forward.

"Community Supported Composting is different," Mr. Brinton has written. "It is not driven at all by economics. It is driven by recognition that the soil needs to be built to provide good, wholesome food. When the system is in place, the consumers know where their food scraps are going, and how they are being used to renew the farmland that grows some of the food they eat."

The CSC model has simple ingredients: plain tools and proven methods. It is something the public can participate in and through which they can give something back to the soil. The basis of the program is small-scale composting: not big, not high-tech, not remote, but simple and straightforward within the community out of which the need arises.

There is an obvious potential to use farms to prepare composting sites for organic residues from local communities. But the local community may rise up in opposition, thinking it will be a huge, noisy, smelly commercial operation—just another landfill. The public must be educated prior to, and as the composting project is brought on line.

Through a CSC a consumer network forms, which embodies the awareness and intent of returning food scraps to the garden via compost made on the same farm where the food was grown. Humus and nutrients then come back into the system from which they were extracted. The process becomes refined again and again over the years.

Based on their years of experience, the Woods End Agricultural Institute emphasizes several key lessons concerning CSC projects.

- Focus on the small-scale, local level—people and institutions which are in a position to do something in their community or neighborhood, without a centralized economic plan being involved.
- Make no attempt to develop the economy of CSC. "This is an important thing," Brinton emphasizes. "In a CSC nobody is trying to sell the product, or barter it, or anything." A CSC should be driven not by standard economics, but rather by the motivation of handling our own wastes wisely, with soil-building as the goal.
- Use a standard small, compact outdoor compost bin made of wood, so that the materials can be easily turned and mixed. This is what Woods End recommends when working with institutions. The standard four-bin composting arrangement can handle a year's supply of food scraps from a medium-sized school or other institution, and keeps the materials separate over time in their stages of development to compost.
- Be sure compost sites are small, and placed according to a plan within the farmer's fields in a neat and tidy manner so they can easily be tended.
- Emphasize the concept of cleanliness, and clean separation as a fundamental part of any CSC. The food scraps must be separated from everything, no mixing with any other garbage, including meat scraps. If the waste streams are co-mingled, farmers cannot make healthy, organic compost to feed the soil.

One of the core elements in a program Woods End developed in New York City is a one-and-a-half gallon biodegradable Food Scrap Bag (see *Appendix F—Resources*). People keep food scraps separate, then just place them in the bag, which has a leak-proof liner. It eliminates the need to keep food scraps in a plastic bag or bucket, where they tend to ferment quickly. When returned to the site, the entire Food Scrap Bag and its contents can be simply added to an active compost pile where it will decompose. The bag has proved popular, and is used in a variety of household and institutional settings.

Tax Supported Farms

Since 1976 the Natick Community Organic Farm in Natick, MA has operated a twenty-two-acre farm about twenty-five miles west of

Boston. It is a loosely defined project on town-owned land, under the direction of the Natick Youth and Human Resources Committee, which the town established to keep kids out of trouble. In 1979 the farm was recognized by the US Department of Health, Education and Welfare as an outstanding alternative program for community youth. As of 1996, the farm was open to all local youngsters who aren't afraid of hard work and long days.

In the summer sixty to seventy students from the town's public schools work the land, shear the sheep, spin the wool into yarn, milk the goats, care for the vegetables and flowers, maintain the farm's passive solar greenhouses, and cut firewood for the following spring's maple sugaring under the guidance of the full-time farm staff. The farm's commercial roadside stand is immensely popular with townspeople, and provides income of over $16,000 a year.

The farm is a tax-supported partnership funded jointly by the town government, the Natick public schools, and local private nonprofit groups. It thrives in a suburban community where only about 7 percent of the land is left in open space.

Besides the students, the farm also hosts women in drug rehabilitation programs, troubled youths who need a job, special-needs students, court-ordered community service volunteers, high school students looking to fulfill a thirty-hour community service requirement, home-schooled students, and people who are clients of the local mental health center.

Farm Director Linda Simpkins says the Natick Community Farm could be a model for many other communities interested in maintaining agriculture and open land. "Anyone in town can come out and use this and as long as they are respectful of it," she says. "So far the community really seems to appreciate that." Many years over 14,000 people visit the farm—students, consumers, educators, nature lovers, and so forth. And every year, for over twenty years, local taxpayers have recognized the value of the farm and voted to spend their public money to support its operations.

CSAs Involving Low-Income or Homeless People

CSAs are well suited to work with low-income or homeless people who may otherwise lack access to fresh produce and information about nutrition. When a group of people work together, as is typical in a CSA, they become aware of each others' capabilities and needs.

The CSA community may well become aware of households that are unable to carry the cost of a full share of the farm's budget, and decide to support those households in a variety of ways. Members in some CSAs elect to pay a small amount extra, so that, for example for every twenty-five households, one full share is carried by the group. That way one family could be provided with food at reduced cost, or no cost. It is also common for CSAs to arrange to have leftover food go to a food bank, a soup kitchen, or directly to families in need—so that nothing the farm produces goes to waste.

The Roxbury Farm CSA in New York state donates its surplus to a large food pantry in New York City, one that operates out of the church where they do their share distribution. They are exploring ways in which they can offer regular shares to a more diverse economic population. Ideas that they hope to incorporate include: food stamps; scholarship shares (partial and full); sliding-scale share prices; and payment over time. While they believe they can support some or all of these options financially, they have not yet worked out the ways to reach new individuals and communicate with them.

Some CSAs have a work-for-a-share option for people with either low or no income. Rather than a monetary contribution, people contribute labor, an hour of labor being valued at minimum wage, or some other level agreed upon in advance. But, training is necessary—people are often unaware of the many kinds and levels of knowledge and skill necessary to do farm work.

Several projects directly link homeless people with CSA gardens or farms. For example, over the course of two years, forty homeless people, a small staff, and many volunteers turned a two-and-a-half acre vacant lot in Santa Cruz, California, into a thriving organic garden. The Homeless Garden Project is intended to offer homeless people an opportunity to move from the margins of society to the center of community activity. For the 130 shareholders in this CSA garden, the project offers a chance for them to invest their food dollars directly into socially and ecologically responsible farming. At the time the project was developing (1991 and 1992) there were over 2,000 homeless people in Santa Cruz County.

The garden itself is a model of sophistication. It employs Alan Chadwick's French-intensive/biodynamic, raised-bed method of gardening. Their intensive practices include crop rotation, companion planting, drip irrigation, high species diversity, and a community composting project.

The operating budget for the garden comes from three sources: one-third from shareholders in the garden; one third from the sale of excess produce and flowers at local farmers markets and restaurants; and one third through seasonal festivals, special events, grant and letter writing, and various innovative direct campaigning activities.

Some workers have found housing through their participation in the garden, though most sleep out in the warmer weather or use local shelters when the weather turns cold. Homeless people who labor in the gardens are compensated at the rate of five to six dollars an hour for about twelve hours of work a week, and they also share in the harvest. Purposeful work at the garden enables many of the homeless to make constructive changes in their lives. One of the Project's most important benefits is the community forum it provides for people, both with and without homes, to come together in ways which open dialogue and build understanding.

Over time the Homeless Garden Project has developed a rich web of relationships with local schools, ranging from the elementary levels to the state university.

Food Banks and CSAs

In addition to providing 400 shareholders with fresh, organic produce, the Food Bank Farm in Hadley, Massachusetts also gives an average 100,000 pounds of fresh food each year to local programs that feed those in need. In doing so, they may have established a model that other CSAs, food banks and food pantries can follow, at least in part. There are many potential benefits for the farm, for the farm members, and for the community at large.

The Food Bank Farm is the first such operation to depend on the CSA model, and to be able to donate half of its harvest to needy people. The farm is part of The Western Massachusetts Food Bank, which was started in 1982 to gather unwanted and surplus food from the government, manufacturers, and distributors, and then to make the food available to programs that feed those in need. From a 500,000 square-foot warehouse in the town of Hatfield, the food bank distributes about two-and-a-half million pounds of food each year to feed approximately 114,000 people. There are hundreds of similar food banks, or food pantries, around North America—though few are able to provide needy people with fresh, organic produce.

After ten years experience, mainly dealing with the kinds of heavily processed foods that are generally available as surplus, the Food Bank staff wanted to find a way to establish control over the food supply that is its stock in trade. So in 1991, after trying some smaller experimental projects, they initiated The Food Bank Farm.

Before the farm became a CSA, it was operated for the purpose of growing food for the Food Bank to pass on to those in need. In those days Farm Director Michael Docter spent time securing grants from foundations for the farm's operating budget. While he was having success, he soon recognized that this was a poor long-term plan: grants were unlikely to keep coming year after year. It was just not a sustainable approach. He began visiting CSAs around the country to see what they were doing, and how they were doing it.

"We saw that we needed to take an entrepreneurial approach to getting fresh, organic produce into the stream of goods distributed by food pantries and agencies," Michael Docter explains. "To do that, we knew we needed innovation and control over our own land."

The farm settled on sixty acres of remarkably rich cropland in nearby Hadley, Massachusetts. With the help of a state law designed to protect rural areas from unchecked development pressures, the Food Bank got the Commonwealth of Massachusetts to buy the land's development rights from the previous owners. Then with a mortgage from the Vermont National Bank's Socially Responsible Banking Fund, they bought the land at a price far below what it would have cost at market value. Typical land in the Pioneer Valley goes for $10,000 an acre, more than any farmer could afford to pay for crop land.

Because the Food Bank and its farm are tax-exempt charitable organizations, with a mission that is easy to understand and support, they were able to secure grant money to pay off the $280,000 mortgage on the farm land in three years.

Organizationally, they set themselves up this way: the Western Massachusetts Food Bank is the private, non-profit owner of the farm's land and tools. Michael Docter, who initiated the project, took on the mantle of Farm Director, and Linda Hildebrand became Harvest Manager. Two or three apprentices, or farm hands, help with the farm work.

To others who might seek to establish something similar, Michael Docter offers clearcut advice: "Secure the money once for capital improvements—for the purchase of land, or tools. Then you will not need to go back seeking funds to support the farm's ongoing operations."

Having secured the financial support for their land purchase at the outset, Food Bank Farm soon stabilized, then began producing a bounty. The income generated from the sale of CSA shares now pays for the entire operating budget of The Food Bank Farm, including the production of food that is distributed through the Food Bank of Western Massachusetts. The farm supports itself and offers a generous and healthful bounty to those members of the community who need the food.

According to an article Michael wrote for the magazine *Growing for Market,* "The farm's ability to attract a sizable shareholder base in a short period of time was strongly aided by its charitable mission. Most people readily understood and responded to the need to feed hungry people. They wanted to help. Membership in the Food Bank Farm CSA is a painless way to do so. For every pound of food that goes to one of the farm's shareholders, the farm gives a pound of food to the local food bank. That is something that could make any CSA shareholder feel good, especially when the cost of their share is no higher than any other CSA, and less expensive by far than buying the food at local natural foods supermarkets."

Docter's economic claim is based upon research. Jack Cooley, a graduate student in the Department of Nutrition at the nearby University of Massachusetts, did a comparison study. He found the cost of a share at The Food Bank Farm ($450), would have cost a consumer $1,150 at a local natural-foods supermarket, and $750 at a regular local supermarket.

However, as Docter observed in his *Growing for Market* article, "Even with the savings, in the first years of operation, as new CSA members found out they were receiving a large quantity of vegetables—many of them unfamiliar and seemingly strange—they abandoned ship. The Food Bank Farm lost over 50 percent of its shareholders each year for two years. Something needed to change. The farm began to diligently survey members, especially departing members, to learn their perception of the farm and its produce." Docter and Hildebrand recognized that with a retention rate of only 50 percent, the farm would soon be bankrupt.

"In addition to everything else it is, this is still a consumer deal," Docter says. "The shareholders have to be happy with what they are getting, or they will leave. So we make sure our shareholders get what they want and get a good deal. Every CSA should survey its shareholders regularly to stay in touch with what they need and want."

According to Docter, the fundamental difference of the Food Bank Farm—the edge that allows them to save shareholders money and also to give a generous supply of fresh produce to the food bank—is the combination of great land and the agricultural experience of the growers. "We work fast," he says, "and we use our resources efficiently." This is the source of the bounty that they are able to offer to the community.

"CSA farmers need to know how to grow a good crop," Docter says. "They also need good vegetable land. Lots of people are attracted to CSA on the basis of ideals, without any real farming experience. Then they go out and try to cultivate a swamp or a rock pit. But without the right varieties, quantities, and qualities, people depart after a year or two, and the CSA can, as many have done already, collapse."

Food Bank Farm has about sixty acres of cultivated land, with up to thirty-five acres in vegetables during the growing season. Their soil is among the best in the world for growing vegetables, the fabled Hadley loam. Without such soil, and without experienced growers, others CSAs are unlikely to match their phenomenal record of giving away a pound of fresh food for every pound they grow for their 400 CSA shareholders. But, Docter says, with good management and growing practices, other CSAs could conceivably give away 20 to 30 percent of their harvest. By doing so, they would strengthen the quality and supply they provide to their shareholders because they are growing a surplus.

"To succeed as a CSA, you must overproduce," Michael Docter believes. "You must grow more than you think you are going to need to compensate for the vagaries of Mother Nature. A lot of wholesale growers wind up giving their food away to someone anyhow, so you might as well give it to someone who needs it."

Harvest Manager Linda Hildebrand grew up on a farm and had lots of experience doing commercial growing for market. She found that she—and most of the other commercial growers she knows—have been enormously frustrated. Their profit margin was exceedingly thin, and then the wholesalers and supermarkets would often reject a crop for seemingly spurious reasons—reasons they often would not state plainly. "With commercial growing the big markets are in control," she says. "The farmers are at their mercy. In that regard, CSA makes all the sense in the world. It restores a measure of control and dignity to the growers."

Docter and Hildebrand say one of their goals since the beginning has been to be economically self-sufficient so that the model of their farm can be replicated elsewhere. In just a few years, they have attained that goal. "If CSAs in general are going to survive long-term," Docter observes, "they have to be good growers and good business people. In another seven years, the CSA farms that are around today will either have figured out both the growing and the business end of things, or they will not be around at all."

To help other CSAs become efficient and financially sustainable, Michael Docter and Linda Hildebrand, in partnership with Dan Kaplan of the nearby Brookfield Farm CSA, have started a consulting organization called CSA Works (see *Appendix F—Resources*). As they explain it, their commitment is to help ensure that the farmers of tomorrow are still farming when tomorrow arrives.

Hybrid CSA

A hybrid CSA is both a CSA and a farm with a traditional mono-, or market crop. This is a way for people who are involved in the commercial agricultural world to move in a different direction without having to completely change overnight what they are doing, and radically altering the basis of their economic security.

Steve Moore, for example, has a large farm north of Los Angeles where he grows lemons and avocados. A few years back he opened part of his land to a vegetable garden. Now, in addition to selling organic lemons and avocados to the wholesale market, he also has a CSA membership. Moore also actively seeks to sell his lemons and avocados to other CSAs by pre-arrangement. Likewise, a biodynamic grain farm in the Midwest sells its grains by pre-arrangement to CSAs and to a cooperative of bakers who buy a portion of the harvest in advance, so they can use that grain to bake their breads.

In the same way, CSAs can stand around one another and buy specific crops from one another for distribution to their members. This is a development that is evolving naturally—a different type of wholesale activity among CSAs.

Gena Nonini raises grapes in what is perhaps the most highly industrialized agricultural area of the world, California's 300-mile long San Joaquin valley. Her involvement with CSA makes a vital difference in her farm's survival, because the margin on her table grapes, when sold just to the wholesale market, is so thin that just one dip in the market price of grapes, and the success of her entire season would be in jeopardy.

Ms. Nonini has sixty acres total, with about forty-five acres in various crops, primarily Thompson seedless grapes, grown with biodynamic methods. She is increasing the diversity of her farm by pulling out several varieties of grapes to plant six-and-a-half acres of citrus. She has also put in a half acre of vegetables, and is interplanting vegetables between the trees.

"I see CSA as an important part of my business" she told Jean Yeager in an article he prepared for *BIODYNAMICS,* the journal of the Biodynamic Association. "I've started to market to other CSAs and that is important. It's hard for me here in the valley because there is no large city nearby in which I can develop a CSA, so I have to work with someone closer to the city. It's good to work with these other folks because we tell one another what we're doing, what we're struggling with, and that helps."

Michael Abelman of Fairfield Gardens in Goleta, CA also operates what might be termed a hybrid operation, with CSA representing just about 20 percent of the farm's income. He also sells freshly picked fruits and vegetables at the Fairview's retail store (40 percent), and at farmers markets in Goleta and nearby coastal cities (25 percent). Via wholesale sales and distribution, they make the final 15 percent of the farm's income. In addition to all of this, in an effort to educate the next generation of suburbanites, he has turned Fairfield Gardens into a lab for school children. Throughout the year, they visit in droves via educational programs at local public schools.

The bottom line for the twelve-acre Fairfield Gardens is that at the end of the year they gross on the average of $350,000, though it costs approximately $349,000 to get there. Every year the farm feeds 300 to 500 families, provides salaries for fifteen full-time employees, provides housing for Abelman and his family and the majority of employees, and makes a small profit.

CSA Script–Alternative Currency

Many CSAs have sought to increase their revenue by selling extra products and produce to shareholders. However, as Michael Docter of The Food Bank Farm wrote in an article for *Growing for Market*, "The various transactional costs—weighing produce, making change, bookkeeping, and so forth—can easily exceed the benefits."

Nonetheless, The Food Bank Farm wanted to offer members some of its surplus, extra products like apples and pears that they were unable to grow themselves but which were grown by other local farmers, and also to offer packaged kits for popular dishes such as pesto, salsa, and borscht (the kits contain all the farm-fresh ingredients for the dish, and a recipe). To make this possible and economically worthwhile, they instituted an alternative currency, or scrip, to streamline the transactions so that the extra sales are truly worth the effort.

At The Food Bank Farm they sell a "Scrip Card" for ten dollars. The card has space for five two dollar punches on it. All extra produce and products are priced in two dollar increments: everything costs either two, four, or six dollars. That way, there is really only one transaction. A farm member buys a ten dollar Scrip, and then uses the card as alternative currency to purchase whatever extras he or she wants. The card is punched, up to five times, to account for the purchases.

Keeping track of Scrip is easy, Michael Docter explains. Each card is numbered and sold in sequential order. The card has a perforated section that is torn off at the time of sale. This stub serves as the farm's receipt. At a glance, they can reckon how many Scrip Cards they have sold, and how much money they have taken in.

When they implemented the Scrip system, The Food Bank Farm found that sales went up dramatically. For example, they went from selling forty bushels of fruit one year, to 150 bushels the next year. The consumers liked the system, and so did the farm's bookkeeper. It consolidated many small, bothersome transactions into one smooth, easy-to-track exchange.

With the Scrip system, The Food Bank Farm is able to offer something special to members who want it, without having to ask all the members to bear the cost. Those who want the extra produce or fruit have easy access to it. The system has worked so well for The Food Bank Farm that they are now planning to expand it by forming a relationship with a small dairy that can provide milk and other dairy products. Meanwhile, other CSAs have adopted similar scrip systems.

So that other CSAs can easily follow their lead, Food Bank Farm has produced a sample of their "In Farms We Trust" Scrip Card on computer in Microsoft Word 2.0 format (see *Appendix F—Resources*).

Membership Turnover–Shareholder Retention

On the average, fifty to seventy-five percent of CSA members rejoin their farm from year to year. Some CSAs, however, have had even higher membership turnover, some as high as seventy percent. This clearly is a serious problem. When membership turnover is high, the community is inherently unstable.

According to exit surveys conducted by a number of CSAs, the principal reasons for drop-outs are: members moving away, not having enough time to cook, or receiving more food than they can use. High turnover rates may also have something to do with unrealistic expectations held by new members. CSA arises out of the realm of ideals, which is what attracts many people; but it operates in the world of soil, and relationships, and a thousand other tangibles. For a CSA to truly succeed, those tangibles must be met with clear will, and well-intentioned effort.

Barring major disasters in the garden, or a change in staff or distribution site, CSAs that are well run typically stabilize after three to four years. The consumers become educated about the farm, about the foods, and about the full scope of what their participation means. The growers become more deeply acquainted with the nature of the CSA undertaking, and the particular community they are part of. After a few years, it is apparent to all whether the people are going to make their CSA work.

To support the process of membership stabilization, CSAs have tried different approaches:

- Some CSAs, recognizing the importance of beauty and color, each year cultivate, harvest, then include bunches of cut flowers with their weekly deliveries. This has tremendous appeal.
- Some CSAs do home delivery of the weekly share, although this option tends to reduce the social connection with farm. Since this is a lot of extra work, it is necessary to charge more for a share.
- Some CSAs have gone to a mix-and-match system to give consumers a wider range of choice, as explained below.

Mix and Match

A common complaint from CSA shareholders is that they get too much of some vegetables that they do not want, and perhaps not enough of those they do want. Several years ago, this complaint was also voiced at The Food Bank Farm in Hadley, Massachusetts.

After the farm lost nearly 55 percent of its members over the first two years, they turned to market research to learn what they were doing wrong. Farm Manager Michael Docter feels this is crucial for all CSAs. "It's really important to know what the farm members are thinking," he explains, "especially those who decide to leave."

When The Food Bank Farm surveyed its members, it found that they were frustrated with the limitations of the CSA form, with a lack of choices. Members complained that the farm did not grow enough of the staples—carrots, corn, broccoli, tomatoes, and lettuce. Docter and Hildebrand responded by planting and harvesting more, but still found a level of frustration among members, who felt locked-in to what the farm grew, rather than having a choice of what they wanted to eat. This is a problem that most other CSAs have also had to contend with in one degree or another.

In response to this reality, the Food Bank Farm established a policy to give members choice and control over the foods they receive. They set up a table at their distribution center for something they called "Mix and Match".

Michael Docter explained his innovation in an article he wrote for

Growing for Market: "This approach is based on the recognition that some people want large quantities of a particular vegetable like radishes or kohlrabi, while others will not want them at all. A CSA can place all of the exotic produce on a table, and give consumers a certain standard-size bag to fill with whatever they like from the table. Then no one is forced to take what they do not want, and those with a special interest in certain vegetables can get a generous supply of what they do want. This gives the consumers more choice and control over their diet. Such an approach is obviously not feasible for CSAs which pre-bag the weekly shares, and then ship them to remote distribution centers.

"There is no need to take what you don't want," Docter wrote. "People come in with a shopping bag and fill it to a pre-determined level with whatever they want. If they are into carrots and lettuce, that's what they take." At first, the farm was concerned that everyone would choose one thing in particular, say carrots, and that they would quickly exhaust the supply of carrots and have large quantities of everything else left over. But they found it didn't work that way. "Some folks do take lots of carrots," Docter says, "but others take lots of eggplant, or peppers, or broccoli. And it all seems to average out in the end."

Consumer response has been overwhelmingly positive, according to Docter and his partner, Harvest Manager Linda Hildebrand. Members feel they are getting more of the things they want, and that as a consequence their families are happier. One member of Food Bank Farm told them that with the new policy it was "like the Berlin Wall coming down." Seeing the level of membership satisfaction this can bring, other farms have followed their lead.

Community Supported Buying Clubs

The consumer households assembled in a CSA make an ideal group for bulk purchases of all kinds from other farms or enterprises, including household cleaning supplies, and other goods and services. In a buying club, the group uses its combined purchasing power to acquire bulk staples—like flour, rice, or corn—and then distributes the bulk goods among the members.

CSAs can link in with buying clubs to establish a low-cost method of purchasing food and other goods directly from wholesale distributors. Through cooperating with other members of the community, or congregation or corporation, members can build community and save between 30 to 40 percent on a wide range of regularly needed kitchen, household and other supplies.

For this kind of service to go smoothly, however, roles and responsibilities need to be clearly defined. Who will seek out and evaluate the goods and services? How will payments be handled—separately or through the CSA treasurer and bank account? What happens if a product or service is not acceptable? As long as these questions are resolved in advance, the CSA can find it has tremendous buying power as a group. (See *Appendix F—Resources.*)

Other Possibilities

- CSAs have also been involved in barter arrangements. Steve Gilman of Ruckytucks Farm CSA in Stillwater, NY, for example, has had an ongoing trade with a family doctor since 1990. The farm offers a full share in exchange for medical checkups for everyone in the family, and necessary shots for the children.
- Condominium Supported Agriculture has not yet been tried, as far we have been able to learn, but it may well be an approach worth exploring. Condominium associations are already-formed communities, with a powerful shared interest through their homes. Some condo associations may see advantages in supporting a local, perhaps even neighboring farmer, to establish a source of clean, local food.

Appendices:
Guides, Samples and Resources

Appendix A

The Basic CSA Concept: Some Suggestions For Getting Started

In plainest terms, Community Supported Agriculture (CSA) is a community-based organization of producers and consumers. The consumers agree to provide direct, up-front support for the local growers who will produce their food. The growers agree to do their best to provide a sufficient quantity and quality of food to meet the needs and expectations of the consumers. Within this general arrangement of associative economy there is room for much variation, depending on the resources and desires of the participants.

If there is a common understanding among people who have been involved with CSA, it is that there is no single formula. Each group that gets started has to assess its own goals, skills and resources, and then proceed from that point. Most CSAs are started by farmers, but many have been started by various community, consumer, and church groups. The decisions any CSA reaches, and the challenges it faces will vary from case to case. Still, though, there are some general suggestions that should be considered by any group or individual starting out.

- Find out whether there is a CSA in your area that you can join (see *Appendix F—Resources* to identify organizations that maintain directories of existing CSAs). If there are no local CSAs, propose the idea to your local garden club, environmental action group, school, PTO, PTA, or church—any group with whom you share common interests.
- Begin by sharing the idea informally with a small number of people. When several people are interested, call a public meeting. Announce it through posters in local businesses, notices in the local newspaper, and elsewhere. Steadily build public awareness of and support for what you are doing. If you get articles about the initiative or the meetings published in the local paper, cut them out and copy them on a flier, then post them in likely spots such a local markets, school bulletins, daycare centers, and so forth.

- Present a clear overview of the possibilities at the first public meeting. Be sure to allow time for general discussion, and include a sign-up sheet to collect the names and addresses of everyone who is interested. Further informational meetings may be necessary.

- Form a core group. For a CSA to get off the ground, it must have a committed core group of five or more people who will regularly attend meetings, and do the work: make copies, place phone calls, execute decisions, and so forth. When idealistic movements like a CSA get started, they often attract many curious people. But of the initial crowds, in general only a small number of people will actually continue to come to the meetings and do the work necessary to organize and operate a CSA. Without a minimum of at least five people, the core group will quickly exhaust itself and may come to feel discouraged.

  The core group is a council of those who grow the food, and those who eat it. There is no universal formula for a core group. Rather it will depend on the intentions, personalities, and skills of the human beings who come together. The principal thing is that they share a commitment to the project. The group does not need to agree on everything—this is not a political coalition with the intention of supporting or thwarting an ideology. It is a cooperative movement to provide good, clean food, and to support the farm and farmers.

  Some CSAs, usually started by farmers, either do not rely on a core group, or rely on it to a minimal extent. The farmers choose to undertake the tasks the core group usually handles.

- Build a list of names and phone numbers, and pass a hat for donations. The core group may well have to donate some stamps and phone calls at the outset, something which most people are willing to do. But if there is no provision for incoming resources from the wider community, the core group may soon feel depleted.

- Develop a clearly defined vision first, supported with specific goals, practical plans, and a true budget for meeting the vision. Community farms will find it helpful also to compose a brief statement of purpose so that what you are doing is clear in your minds, and can be easily and reliably communicated to others. It may take many meetings to define the vision and set the plan, and the plan may change over time, but these are essential and healthy parts of the process. If someone attempts to establish a CSA as if it were a rigid franchise concept, then other people will have been denied a chance to contribute to and build the idea. Consequently, they will have no sense of belonging, and perhaps a low level of commitment.

According to experienced community organizers, a purpose or mission statement should be brief, possibly even as short as one sentence.

Spring Hill Community Farm in Prairie Farm, Wisconsin, had its first CSA season in 1992. During the winter of 1992–93 they developed this statement: "The mission of Spring Hill Community Farm is to bring people and land together in a practical, enduring, life-giving community. We do this by growing and distributing food in a way that is just and sustainable; that helps us develop skills and knowledge of land and community stewardship; and that connects rural and urban people to the land and each other in a way that celebrates life and harmony with the earth."

Four Eagles Garden in Ashland, Oregon has a one-sentence mission statement: "Helping each other to live in a good way and ensure a future of healthy land, water and food."

- Foster a democratic process based on consensus. A CSA may want to rotate the chairmanship from meeting to meeting in the beginning. Having a different facilitator for different meetings, ensures that the power or authority to make decisions does not become vested in the hands of one or two members, but is instead widely shared. Such democratic processes typically strengthen the sense of membership in the group, and help to prevent individuals from overextending themselves. In some CSA core groups each new issue is discussed. But if the group cannot come to agreement right away, the issue will be tabled. Some issues will either solve themselves, become irrelevant, or self-destruct. But if the issue does come up again, the possibility of agreement is often enhanced just via the passage of time.
- Have a well-defined agenda for each meeting and stick to it. Whoever is going to chair one of the organizational meetings should, at the outset of the meeting, define what the agenda is and the order in which items will be discussed. These straightforward business procedures help ensure that the creative energy of the group members is clearly focused. If meetings are short—say one two-hour meeting a month (including refreshments and socializing), and also tightly focused, the core group will be more likely to maintain its drive and effectiveness.
- Share responsibility. The more members who are carrying the weight of the CSA and participating in decisions and operating activities through the core group, the greater the chances of success. Some core

groups are as small as four or five members, and some as large as twenty members—with different roles and responsibilities spelled out in "job descriptions."

- Be systematic. The CSA concept arises out of the realm of ideals, but it must be based in practical and efficient business systems. Computers, for instance, can be a tremendous help in keeping track of members, finances, and crop estimates. While farmers and gardeners may lack such skills, other members of the group may be able to offer assistance with computers, businesses practices, and other technical skills.

- Establish the administrative dimension of the CSA, the tasks which will be undertaken by the farmer or the core group: bookkeeping, mailings, phone communications (which can be substantial), and the preparation of printed materials such as announcements, the prospectus, and the newsletter.

- Make the basic decisions of how many people to employ, what jobs the employees should cover, compensation and benefits, and what work is to be done by volunteers. Clearly defined work responsibilities are important to prevent confusion and burnout.

- Bear in mind that while many CSAs have been started by farmers, many have also been started without farmers or gardeners. These groups have simply organized themselves as a community with the clear intention of supporting a CSA, then sought out the personnel to grow their food.

- Appreciate that the farmer or gardener need not be full time. Some CSA projects have as many as four gardeners to provide up to 500 shares, each gardener averaging twenty to thirty hours of work a week. This arrangement—which can be varied to meet the needs of the group and the gardeners—allows the workers to pursue other interests or occupations simultaneously. It also makes it possible for the gardeners to cover for each other if pressing business, such as family births, weddings or funerals, call them away from the fields.

- Determine in general what kinds of vegetables, and how much of each kind, consumers are interested in. In this way the farmer or gardener learns the needs of the community before planning the season. This helps overcome the element of chance or uncertainty in the existing market, and can eliminate considerable waste.

Once the farmers have an idea of what the community needs are, they can draft an annual farm budget, taking into consideration all the costs associated with planting, cultivating, and harvesting the produce, along with the needs of all those who agree to provide the service, and their dependents (see *Appendix C*). Be sure to also figure in the cost of phone calls, and the printing and mailing of materials such as the prospectus and newsletter. Dialogue and negotiation take place around the needs of the consumers and farmers, and the costs of production to meet those needs. From this dialogue and budget, the price per share is established.

• Establish the share price after determining the budget. Plan the budget as accurately as possible; you will be paying the bills. Perhaps the number one reason that some CSAs have gone out of business is that they failed to budget properly, or to ask a sufficient sum in exchange for a share. Some groups supplement the share prices by selling investment shares, or "contributing member" shares to people and organizations who want to support the idea, such as local churches, community groups, or even banks. Their share in the harvest can be donated to the local food bank, or other hunger programs. This allows both the giving and the receiving organizations to be directly linked with the CSA.

• Prepare a prospectus—a one-page document that states plainly what each CSA shareholder is expected to contribute financially, or through labor, and specifically what each shareholder can expect to receive if the season goes more or less according to plan. This is the agreement, or contract, between the member households and the CSA they are part of. (See *Appendix D* for a sample prospectus).

• Decide on your distribution method, and whether the CSA will require work of the members. An attractive site for harvest distribution—preferably on the farm—helps maintain community pride and morale. Work contributions can be all volunteer, a set number of hours for a reduced share price, or even the opportunity to work off the entire share. Jobs typically include helping with planting, weeding, harvesting, creating posters, brochures, or a newsletter, or even doing the accounting. Some of these may be paid, part-time positions—the group can decide. Bear in mind that most volunteers will need training and guidance from the active farmers.

• Consider whether the CSA should make accommodations for working members, who may perform networking tasks such as phone calls, or preparing the newsletter, or even with farm chores such as weeding or harvesting. In exchange for their labor, working members can, for example, make half-payments. Thus, if a full share in the farm's output was valued at $400, the working members would contribute some labor, and pay just $200. This must be planned for as the farm budget is prepared. Determine how many half-memberships can be handled financially, if any, and then offer the working memberships to the general CSA membership—interviewing those who are interested to determine their interests, skills, and availability. Time must be spent with each new worker and each new task, to show how, when and where. Follow-up is also necessary. Not many people possess the variety of skills needed for farming.

- Strive to establish a land base, either with a long-term lease or through a permanent land trust (see *Appendix B*). While the acquisition of land may take a number of years, it is essential for the ultimate success of a community farm. If a CSA is forced to rent or lease land for growing from year to year, then those who work the land will have less motivation for making long-term improvements and investments in upgrading the quality of the soil and the physical infrastructure. Additionally, the general membership of the CSA will have no focal point on the land for its energies, something that is essential if the CSA is to maintain a sense of stewardship with the earth.

- Consider using computers to lighten the work load. While a particular farmer may not have the equipment or expertise for this, often a member of the core group, or a general member of the farm community, will. Articles for the newsletter, surveys, database directory listings, financial spreadsheets, letters to resource providers, and so forth can all be created and stored in files and folders. It can also be used for application forms, recipe sheets, food guide materials, membership roles, mailing lists, and other organizational elements. Some CSAs use computers to keep track of planting schedules, harvest quantities, crop rotations, and field plans. And some CSAs have even established homepages on the World Wide Web to disseminate public information about who they are and what they are doing. This reduces the need for someone to sit at a desk to handle inquiries, and mail out responses.

- Publish a newsletter. Most existing CSAs have newsletters, generally on a quarterly basis, though some are monthly, and yet others are weekly or bi-weekly during the growing season. The newsletter serves as an essential communications link to keep members informed of the decisions and developments that concern them, and also serves as a means of educating member families about what they are receiving, and how to prepare or store the foods. Without regular news of the CSA, members may lose touch and then lose interest.

The newsletter can be put together by a member of the core group, or a general member of the CSA who has time and experience.

- Engage in shareholder education on how to use the food—because most of us have been part of a processed economy, fresh vegetables may be unknown or little understood. Newsletter articles and recipes have proved enormously helpful for successful CSAs. Food preparation and storage classes conducted by experienced members of a CSA have also proved their value. (See *Appendix F—Resources*).

- Contact other CSAs, or CSA organizations such as the Biodynamic Farming and Gardening Association. Let them know what you are doing. Find out what resources they make available to farmers and farm groups. There is no need to re-invent the wheel; you can benefit from learning how other groups have succeeded, and how they have failed.

- Survey the shareholders regularly, especially those who have chosen to depart, to learn their view of the farm, and to stay in touch with the quantities and qualities of crops they want. The survey can be formal or informal. This is a central issue. CSAs that fail to understand and thereby fail to meet shareholder needs and wants are often the ones that fail.

- Expect change. Every CSA—as with every human institution—experiences an ongoing process of change as people come and go over the years. It is inevitable. Resources and group capabilities and intentions are also likely to change.

- Be patient. It takes time to make any community undertaking work. If a community farm is going to be successful, it will have to evolve gradually over time. But if it's worth doing, it's worth taking the time to develop it. Most successful CSAs have taken three to five years of transition to establish the mature model that meets the needs of local growers and consumers; even after that, change is an ongoing part of the process.

Appendix B

Gaining Ground: How CSAs Can Acquire, Hold and Pass On Land
Chuck Matthei

While the first decade of Community Supported Agriculture in the United States has been impressive and prospects look bright, the majority of CSA farmers are still on shaky ground. Many are working borrowed or rented land, without long-term leases; some are trying to purchase land at market prices far higher than the productive agricultural values; and even those who own land are trying to figure out how to provide for their retirement and their heirs, yet insure the continuation of farming in future generations.

Nothing is more important than secure land tenure and reliable sources of financing. The lack of it poses a threat to existing CSAs and a barrier to would-be farmers. But the current problems can be solved by mobilizing the unique community that CSAs have created, and in the process CSAs can make a valuable contribution to other farmers and the larger community as well.

The reality is that most CSAs will not be able to obtain sufficient financing from conventional sources, and many will need substantial discounts or subsidies in any event. The key to finding these resources lies in distinguishing the essential personal interests in farm properties from the inherent public interests. Defining and protecting the public interests legitimizes the application of charitable and public funds to a land purchase, and thereby assures affordable access and full opportunity to the farmer. In most cases, this is achieved by establishing a relationship and dividing the property interests between the farmer and a nonprofit land trust, with the land trust serving as steward of the public interest.

Finding Your Place
Most CSA farmers begin as apprentices, and many spend their first few years as independent producers in a temporary location. But eventually

232   Chuck Matthei is President of the Equity Trust, Inc.

each will need a secure site in which to invest many years of labor and a substantial amount of capital, building a sustainable farm operation.

Choose your land carefully, paying more attention to productivity than panoramic views. With the experiences of others in mind, be realistic about size, soil quality, availability of water, and the various improvements that a successful farm will require. Be mindful, too, of the terms of access offered to you, including the length of the lease, permitted land uses, and the possibility of credit for improvements you might make. Whenever you enter into a lease, ask for a right of refusal or purchase option in the event that the owner puts the land up for sale. Consider the market value of the property and the level of interest and assistance that you might find in that community, should you have to purchase the land to secure your tenure.

There are many ways to acquire land at below-market pries. Look for motivated sellers who may be willing to take less than market value to insure that the land they love remains a working farm—or others who might be attracted by the tax benefits that come with a bargain sale to a nonprofit organization. (The latter are usually landowners whose property values have appreciated significantly, and who have enough income or estate value to use the charitable deductions, which can be spread over several tax years.) Some elderly landowners may be willing to reduce the sale price if you will allow them to remain in their homes, and perhaps provide some personal services to them after the land transfer.

Local clergy, attorneys, investment advisors, and sympathetic realtors may be in a position to know which landowners fit these descriptions. Land trusts and town officials may also know, and sometimes the land trust, local government, or other institutions may already own land that is suitable for farming and available for lease or purchase. The community-building aspect of CSA does not stop with the recruitment of members; it requires relationships and credibility throughout the community.

Forming a Relationship with a Land Trust

There are two families of land trusts in this country, conservation land trusts and community land trusts, with significant similarities and differences between them. Typically, they are both nonprofit corporations (not legal trusts), locally based and democratically structured, but there are variations. The same legal and financial tools are available to both groups, and they are capable of a variety of tenure arrangements, but they may also have accustomed practices, and limitations of knowledge and experience.

The conservation trusts currently number about 1200, with a membership that is primarily middle and upper-class. Traditionally, they have been devoted to open space preservation, but about 10 percent now express an interest in active farm and forest lands, with a few, like the Vermont Land Trust and the Marin Agricultural Land Trust in California, specializing in working lands. Conservation trusts will most often hold an easement on the land, leaving the fee interest or title in the name of the farmer.

In contrast, there are only 120 community land trusts and they are primarily located in lower income communities. Most have been established to provide the essential benefits of ownership to those who are excluded from the real estate market. Many are in urban areas and few have farm holdings, but they should be receptive to a CSAs proposal. Community land trusts usually retain the land title and offer residents a lifetime, inheritable lease; lessees may own the improvements and build equity by their personal investment of capital and labor. The land lease includes an option on these improvements to enable the trust to control transfer and maintain affordable access for succeeding farmers. This difference between the individual holding title and the land trust doing so may seem significant, but in fact it may not be. The character of the land tenure relationship—the distribution of rights and responsibilities—is actually determined by the details of the legal agreements, rather than the type of instrument used.

While the number of land trusts, of both kinds, interested in farmland preservation is steadily growing, the negotiation of terms will be a learning experience for all concerned. The relationship between a farmer and a land trust is a very important and long-lasting one, and it will take time and patience to develop. There may be some bumps along the road, but it is usually better to persevere with an existing local trust than to try to create a new one for a single purpose.

When you encounter the local land trust, look at its stated purposes and history of program activity, its board of directors and membership, its legal and financial condition and its management systems. Make sure that it is a sincere, competent, active organization. If it is, make the effort to build an effective working relationship, turning to others for examples and assistance along the way. If this is not possible or no local land trust exists, you may consider organizing a new one or, perhaps more likely for the near term, utilizing an outside organization with similar purposes as an interim steward.

Balancing Individual and Community Interests

Most of us are accustomed to regarding property as a legal formulation or a market calculation, but it is more helpful to envision it as a web of relationships. The leases, easements, and other documents used to secure land for CSAs should carefully define and equitably balance the legitimate interests of all involved parties.

There is no single right way to do this, and the law is quite flexible. In most cases, you can strike whatever balance seems fairest to you and your partners. A simple exercise might be useful in making this determination. Do it alone, with your CSA core group, and with your land trust partner. Let this be not a dry legal negotiation, but a creative educational and social experience with implications for others as well.

Thinking of the land that you hope to acquire—its natural features, potential uses, and carrying capacities—make a grid. On one side, list all of the interested parties. They may have different kinds of interests and different levels of interest, but they have some legitimate interest in that land. You will find that they can be grouped into four categories: one or more individuals who personally use the land for residence, farming, or some other purposes; the community in the form of abutting neighbors, local government, and the CSA membership; the land itself, and the plants and animals with which we share it; and the next generation.

The other set of coordinates will be the three dimensions of property: environmental, social, and economic. How should the specific rights and responsibilities in each of these dimensions be distributed among the various interested parties. What land uses shall be permitted, required, or prohibited for each? Who shall have access, exclusive or shared, and who shall participate in governance, in the different kinds of decisions that will be made regarding the land? Who contributes to property value over time, and how shall equity be allocated? In all of these areas, think about what you hope to achieve and what else might occur, making provisions for unexpected or even unwelcome contingencies.

With this matrix in hand, you will be ready to begin crafting the necessary legal agreements. The resulting documents, in part, will define: parties to the agreement, including the farmer(s), spouse(s), the land trust, and perhaps others; the land and resources being allocated or withheld from each party, including boundary lines, timber, mineral and water rights, and specifications for private use, shared use, and perhaps public access; the amounts of the lease fees or purchase prices, and the responsibilities for taxes, insurance, maintenance and monitoring; permitted and restricted land uses (residential, agricultural, educational, commercial, etc.) and practices; ownership of the land and improvements, the right to make additional investments and improvements, permitted mortgaging, and the land trust's option upon sale by the farmer; provisions for continued occupancy, subleasing, inheritance, conflict resolution in the event of disagreements between the parties, and more.

Financing the Purchase

If the land is to be owned by the land trust and leased to the farmer, it may be acquired as a (tax-deductible) charitable gift or purchased entirely with gift funds, but the farmer will pay a reasonable lease fee for the use of it. If the farmer will own the land and the land trust hold a conservation easement, they will each pay their proportionate share of the property value as determined by appraisal.

Again, charitable or public resources may be used for the acquisition and maintenance of public interests in property; private property interests, just like the operating costs of the farm, must be paid with private funds. Some CSA farmers have considered restructuring their farms as charitable organizations and a few CSAs, like Quail Hill in Amagansett, NY, and the Food Bank Farm in Hadley, Massachusetts, are currently operating as programs of nonprofit corporations (the Peconic Land Trust and the Western Massachusetts Food Bank, respectively), with the farmers as employees. But this approach may only be feasible when the organization has a broader array of charitable, educational, or conservation activities. The Internal Revenue Service does not recognize farming, as such, as an exempt activity. In most cases, the farm business of the CSA will remain a private enterprise, owned by the farmer or, conceivably, by the farmer and members as a cooperative, with the role of the nonprofit partner limited to stewardship of the public interests in the land.

Before seeking financing for land acquisition, a CSA must formulate a realistic, multi-year business plan, detailing projected income and expenses, anticipating growth, providing for contingencies, and identifying the amount remaining for debt service. On this basis, you may be able to approach conventional lending institutions for a portion of the purchase price. They will require a down payment and expect to see evidence of the farm's capacity and cash flows. They may also ask for co-signers or guarantors of the loan, a role that some friends, family, and

CSA members may be willing to play (with the option of limiting their personal liability and sharing the risk among several supporters).

Some states and municipalities have provided grants or loans to land trusts for farm acquisitions, through established programs or special appropriations. Local foundations, other institutions, and even businesses have also contributed. Typically, the financing package comes from multiple sources and includes a combination of gifts, loans, and perhaps even proceeds from the sale of partial interests to other parties.

It is useful to acquaint yourself with these institutions and programs well in advance of your need for funds, but institutions are not the only potential source of financing for CSAs. The members themselves, and their own friends and associates may be the most important financial resource. They are already interested, involved and motivated. Their annual share payments provide the operating budget for the farm and yield direct personal benefits. Presented with a realistic proposal, they may also be willing to make charitable gifts and socially-responsible investments, to secure the farm for future generations and preserve the character and quality of life of the surrounding community.

Contributions may go directly to the land trust. Many investors, however, may prefer to make their loans through a qualified intermediary such as a community development loan fund. Multiple investors make loans to these funds, for general purposes or designated for specific projects. The funds aggregate the moneys and provide financing to CSAs and a variety of other community development and conservation projects. For investors, the intermediary assumes the responsibilities of analysis, administration, and monitoring, and offers the greater security of its diversified portfolio, loss reserves, and net worth. Borrowers benefit by dealing with a single lender, one with experience, technical assistance capabilities, and additional capital if needed.

Setting an Example

A growing number of CSAs are following the paths outlined above, breaking new ground and enlarging the opportunity for others as they go. Stephen and Gloria Decater had been farming for nearly twenty years—initially as market gardeners and then as Live Power Community Farm, California's oldest continuously operating CSA—when they realized in 1991 that they had to purchase the land. Throughout this period, it had generously been made available by Richard Wilson, a sympathetic landowner, rancher, and Director of the California Department of Forestry and Fire Protection. But with the need for substantial additional investments in the buildings and the eventual prospect of intergenerational transfers, the time had come to formally secure their tenure.

They negotiated a purchase agreement. Like most small farmers, however, they couldn't afford to pay the full market value with only farm income. So they turned to their core group of members for assistance and together began to explore their options. After two years of research and interviews with farmland preservation groups around the country, they chose the model in which they would personally obtain financing for the agricultural value of the property, while members and others made charitable gifts for the purchase of an easement by a land trust.

As they discussed their mutual goals, they decided that a conventional conservation easement, simply preventing inappropriate development and protecting critical natural features of the land, would not be good enough. In a newsletter report to the general membership, Stephen wrote, "Socially and ecologically responsible agriculture also requires socially and economically responsible land ownership. [If] equity and stewardship of the land are shared by the community and the individual farmers. . . we can ensure that the land will remain in farming use and permanently affordable to farmers."

With the help of attorneys and other advisors, they crafted an easement and purchase option for the land trust that not only provides for environmental protection, but requires that the land be continually farmed, by resident farmers, using sustainable methods—and limits the price, when it transfers from one farmer to the next, to no more than the productive farm value. On this basis, a second appraisal was made: the market value of Live Power Community Farm was $150,000, but the restricted value was $69,000. (The additional provisions of this easement substantially increased the amount of charitable funds that could be applied to the purchase. Covelo, California, where the farm is located, already has large-lot zoning and a conventional easement would have had relatively little impact on the appraisal, because the possibility of future sales to estate buyers and other non-farmers would still be open. Dedicating the property to agriculture and limiting the transfer price, however, removed all of the speculative element and made the farm affordable for the Decaters and future generations.)

In a remarkable effort, the group did succeed in raising $90,000 for the easement ($81,000, the difference between farm value and market value) and related costs. Yet another hurdle remained. The plan required the participation of a nonprofit partner. There was no local land trust in the immediate area and the nearest one, though ostensibly interested and appreciative of Live Power Farm's intent, was hesitant to take on the practical responsibilities entailed in the social and economic preservation of agricultural lands.

At this point, the Decaters turned again to the Equity Trust, Inc., a small national organization with a program of land reform and community development finance, which had already been a source of advice and technical assistance. The Equity Trust serves community land trusts and conservation projects in various parts of this country and occasionally abroad, and it agreed to play a surrogate land-banking role until local stewardship became available.

Finally, in May of 1995, the closing took place. As Richard Wilson said at the celebration: "Along the way, we learned some lessons that may be relevant for others. It takes patience and fortitude. This work is important social reform and it cuts against the grain of existing expectations and arrangements in the marketplace. . . . [But] this farm is a working example of how sustainable agriculture can succeed. It's an important center of education and training. It's the center of a vibrant community that links Covelo to the city and provides the city with a vital contact to the real world of nature and its limits. And it's the home for a wonderful family who are committed to the land and to Round Valley."

Several factors contributed to this achievement. Live Power Community Farm was well established and the Decaters' personal dedication and abilities were well known. The core group included people with significant legal, financial and fundraising skills, and they were willing to devote a great deal of time. And half of the 140 member families live in San Francisco, one of the most receptive and affluent environments for such an appeal.

Other CSAs may have more or less difficulty in their own land acquisition and fundraising efforts, but this experience is not unique. In Massachusetts, the Food Bank Farm paid off the mortgage on its farm fields in just three years, with charitable gifts. (The original price was reduced by the sale of an easement to the state Department of Food and Agriculture; the compound of buildings was acquired by an interested philanthropist and leased to the Food Bank and farmers, with an intention to eventually transfer ownership to a land stewardship organization.) Philadelphia Farm, in Osceola, Wisconsin, was also purchased with the help of charitable gifts, and Fairview Gardens, in Goleta, California, is now (1997) in the midst of its own capital campaign, with the Equity Trust providing intermediary services.

The same principles can also be applied to intergenerational transfers, as an expression of the owners dedication (and, if useful, for the charitable gift deduction and estate tax benefits that result). Roxbury Farm, in Claverack, New York (see *Example 7*), has been in the same family for several generations. Two of the nine children are involved in the CSA operation. After careful consideration and a series of family meetings, the parents decided to transfer the prime agricultural land to the farmers, reserve a small tract of non-agricultural land for each of the other children to enable them to return if they choose, and donate a conservation easement to a local land trust. They met the needs of each family member and fulfilled their common commitment to the land, the farm, and the well-being of the surrounding community. Now they are working with the land trust, neighboring owners, and prospective farmers on a broader strategy for land conservation and agricultural revitalization.

Seeds for a Future Harvest

CSAs like these are defining the principles and perfecting the instruments for a more effective approach to agricultural conservation, for farms of every kind. The essential element in all of their stories—and many others like them—is the willingness of the farmers to balance their own individual interests with the common good and to address all three dimensions of property: environmental, social and economic. While most conservation programs throughout the country, both public and private, are still using tools that protect only the land, these CSAs are setting a higher standard, striving to preserve farmland, family farmers, and rural communities as well. They have taken to heart the warning of Aldo Leopold, a half-century ago, that we abuse land because we regard it as a commodity belonging to us, and learned instead to see it as a community to which we belong.

Along the way CSAs are educating and inspiring, discovering allies, and forging partnerships. The Vermont Land Trust, a national leader in the conservation field, is now experimenting with shared equity models for family farms, and the Commonwealth of Massachusetts has revised the easements used in its Agricultural Preservation Restriction program to require continuing agricultural use. The quasi-public Vermont Housing and Conservation Board holds a statutory right-of-refusal on any farm that has received property tax considerations or other state subsidies, before it can be sold and removed from production, and provides financing to local land trusts through a fund capitalized by legislative appropriations. Achieving our larger goals will involve CSAs with many others in a combination of individual initiatives, local organizing and institutional development, and public policy reforms.

Significantly, the relevance of these efforts goes well beyond rural America, for land is not only essential to farmers, but the foundation for virtually all social and economic activity. Some of the same market forces that are keeping prospective farmers from the land are affecting urban areas as well, where community land trusts in cities large and small are responding to the needs for affordable housing, open space, and facilities for small businesses and human services.

No issues are more important, or potentially controversial, than property rights. In national, state, and local arenas, they stand at the center of the political stage as the focus of highly polarized debates. Public and private interests are typically portrayed as quite distinct and even antagonistic; conceptual vocabulary is simplistic, and constructive alternatives are lacking. Although we define the word equity both as a financial interest in property and as a moral principle of fairness, all too often it seems that we have forgotten the necessary relationship between the two.

Land reform may be foreign to most Americans, but our need for it can only grow. Here and abroad, we are confronted with expanding populations, resources limited by supply or the costs and consequences of extraction, and a universal demand for inclusion. The conclusion is inescapable, even if the path is still unclear. All human beings need food, shelter, an opportunity for productive labor, and recognition of their importance to the community—and all of these are dependent upon good stewardship and equitable distribution of the land.

With creativity, commitment, and community participation, the CSAs of today can offer a legacy of secure, productive, and affordable land to the farmers of tomorrow and make vital contributions to a larger process of social and economic reform. The ultimate success of community supported agriculture depends upon it and others will greatly benefit. It's a formidable challenge, to be sure, but it is also a remarkable opportunity, a practical possibility, and even a sacred trust.

Land Tenure Resources

Recognizing the widespread need for land tenure counseling and financing among CSAs—and the enormous potential of CSA members across the country—the Equity Trust, Inc., a nonprofit organization with an innovative program of land reform and community development finance, has established the Fund for Community Supported Agriculture. The

Equity Trust offers information, technical assistance, and sample documents through conference presentations, workshops, telephone consultations and site visits. A pamphlet with a more detailed treatment of the subjects introduced above is now in production.

The Fund for Community Supported Agriculture invites socially-responsible investments and contributions from CSA members and others, on a variety of terms, and provides financing for land acquisition and capital improvements. It is administered with the advice and participation of CSA farmers and members. The Equity Trust facilitates the development of relationships between CSAs and local land trusts, and may serve as a surrogate when local stewardship is not available. For more information, contact the Equity Trust, 539 Beach Pond Road, Voluntown, CT 06384; phone/fax: 860-376-6174.

Informational and technical materials on conservation land trusts are also available from the Land Trust Alliance (1319 F St. NW, Washington, DC 20004; phone: 202-638-4725); similar resources on community land trusts may be sought from the Institute for Community Economics (57 School Street, Springfield, MA, 01105-1331, phone: 413-746-8660). Both organizations offer technical assistance and host periodic conferences and workshops for their constituencies. The American Farmland Trust also addresses practical issues of farmland preservation and public policy reform.

The Biodynamic Farming and Gardening Association (PO Box 550, Kimberton, PA 19442; phone: 1-800-516-7797) has a variety of publications, including some information on land tenure, and a national referral service to locate other CSAs.

CSA West, a program of the Community Alliance with Family Farmers (PO Box 363, Davis, CA 95617; phone: 916-756-8518) has produced a case study of the Live Power Community Farm experience discussed in this appendix.

Most of the institutional sources of financing which may be receptive to CSAs are local or regional. For example, in addition to the Vermont Housing and Conservation Trust Fund, there are Vermont National Banks Socially Responsible Banking Fund; the nonprofit Vermont Community Loan Fund (a member of the National Association of Community Development Loan Funds); and a fund operated by the state chapter of the Northeast Organic Farmers Association. These organizations provide support and resources specifically to Vermont groups; therefore, you will have to research the possibilities in the state where you are located. One place to begin a search for state and region-specific programs is The National Association of Community Development Loans, 924 Cherry St. Philadelphia, PA 19107; phone: 215-923-4754.

The Trust for Public Land (116 New Mongomery St., San Francisco, CA 94105; phone: 415-495-4014) will sometimes negotiate bargain sales, purchasing and holding properties for local land trusts until they are able to acquire them. Again, you should talk with the land trusts, community developers, and traditional farm organizations in your own area to identify prospective public, private, and charitable lenders and funders.

The American Farmland Trust (920 N Street NW–Suite 400, Washington, DC 20036; phone: 202-659-5170) is a national, non-profit group primarily dedicated to the protection of the nation's farmland. Over the years it has responded to hundreds of state and local requests to help select and implement farmland protection programs, often using the methods standard to land conservation trusts: adoption of conservation easements, special districts for agricultural activity, and programs to compensate landowners for voluntarily relinquishing the development rights to their land.

Appendix C

Sample Budgets

Food Bank Farm Operations Budget—1996

FOOD BANK FARM OPERATIONS BUDGET		1996
INCOME FROM OPERATIONS		Yr. End
		Projected
	FARM SHARES	151,570
	DESIGNATED DONATIONS	3,853
	DESIGNATED GRANTS	0
	OTHER INCOME--FARM	9,297
	INTEREST INCOME-FARM	0
	DIVIDEND INCOME	12
	REVENUE CHILI-FARM	0
	DONATIONS IN KIND	0
	GAIN ON SLE OF A	465
	MISCELLANEOUS INCOME'	0
	TOTAL INCOME OPERATIONS	165,197
OPERATING EXPENSES		
	SALARIES-FARM	75,113
	CONSULTANTS-FARM	0
	WORK-STUDY EXPENSE	0
	PAYROLL TAXES--FARM	5,746
	MED DENT LIFE FARM	6,500
	PENSION--FARM	1,813
	WC INSURANCE FARM	5,172
	OFFICE SUPPLIES FARM	870
	POSTAGE--FARM	1,700
	PRINTING COPYING	1,600
	TELEPHONE	1,400
	UTILITIES	1,650
	PROGRAM SUPPLIES	3,729
	REPAIR MAINTENANCE	4,611
	FREIGHT	
	VEHICLE EXPENSE	819
	MILEAGE TRAVEL	100
	CONFERENCE TRAINING	25
	ADVERTISING MARKTG	
	SMALL EQUIPMENT	745
	RENT	3,636
	SEEDS	6,017
	TRANSPLANTS	3,000
	EQUIPMENT RENTAL	300
	FOOD PURCHASES	8,309
	FERTILIZER	7,100
	FIELD PREPARATION	875
	ACCOUNTING	32
	INSURANCE	467
	FEES LICENCES SUB	263
	VOLUNTEER	168
	SPECIAL EVENTS	
	BAD DEBTS	310
	COGS CHILI	
	LOSS OR DAMAGE TO	
	DEPRECIATION	3,073
	TOTAL OPERATING EXPENSE	144,943
GROSS PROFIT (LOSS)		20,254

Forty Acres and Ewe Budget Report (1/1/97–12/31/97)

Category	1/1/97 Actual	- Budget	12/31/97 Diff
Inflows			
Balance	0.00	34.82	-34.82
Dividend	18.59	0.00	18.59
Gift Received:			
Harvest Feast	0.00	500.00	-500.00
Misc.	25.00	0.00	25.00
Pancake Supper	0.00	500.00	-500.00
Research	125.00	0.00	125.00
Total Gift Received	150.00	1,000.00	-850.00
Gross Sales:			
Institutional	4,000.00	0.00	4,000.00
Lamb & Beef	0.00	1,000.00	-1,000.00
Producer's Pool	0.00	7,000.00	-7,000.00
Restaurants	0.00	8,000.00	-8,000.00
Vegetables	3,695.69	40,000.00	-36,304.31
Total Gross Sales	7,695.69	56,000.00	-48,304.31
Total Inflows	**7,864.28**	**57,034.82**	**-49,170.54**
Outflows			
Auto	90.58	300.00	-209.42
Bank Charges	5.20	150.00	-144.80
Capitalization:			
Payment	0.00	3,000.00	-3,000.00
Capitalization - Other	225.00	0.00	225.00
Total Capitalization	225.00	3,000.00	-2,775.00
Custom Hire:			
Butchering	0.00	600.00	-600.00
Child Care	16.00	1,500.00	-1,484.00
Garden Help	0.00	2,500.00	-2,500.00
Total Custom Hire	16.00	4,600.00	-4,584.00
Delivery	0.00	1,000.00	-1,000.00
Education	34.70	500.00	-465.30
Feed Purchased	91.84	800.00	-708.16
Fert, Lime	0.00	1,000.00	-1,000.00
Fuel	37.45	500.00	-462.55
Insurance	295.94	600.00	-304.06
Interest Paid	1.23	20.00	-18.77
Office:			
Ads	500.00	0.00	500.00
Postage	13.17	250.00	-236.83
Printing	0.00	400.00	-400.00
Supplies	0.00	100.00	-100.00
Total Office	513.17	750.00	-236.83
Rent:			
Land	551.88	3,200.00	-2,648.12
Mach, Equip.	0.00	1,200.00	-1,200.00
Total Rent	551.88	4,400.00	-3,848.12

Forty Acres and Ewe Budget Report (1/1/97–12/31/97)

Category	1/1/97 Actual	- Budget	12/31/97 Diff
Repairs	30.00	2,000.00	-1,970.00
Salary:			
Animal Purchase	0.00	75.00	-75.00
Chicken Feed	0.00	500.00	-500.00
Custom Hire	0.00	100.00	-100.00
Salary	1,895.40	30,089.82	-28,194.42
Salary - Other	2,650.00	0.00	2,650.00
Total Salary	4,545.40	30,764.82	-26,219.42
Seeds & Plants	712.14	1,500.00	-787.86
Storage	0.00	100.00	-100.00
Supplies:			
Fencing	0.00	300.00	-300.00
Garden	0.00	400.00	-400.00
Greenhouse	270.52	750.00	-479.48
Misc.	0.00	500.00	-500.00
Tools	0.00	200.00	-200.00
Total Supplies	270.52	2,150.00	-1,879.48
Telephone	-44.81	700.00	-744.81
Truck:			
Fuel	65.77	600.00	-534.23
Service	0.00	600.00	-600.00
Total Truck	65.77	1,200.00	-1,134.23
Utilities	140.17	1,000.00	-859.83
Outflows - Other	0.00	0.00	0.00
Total Outflows	**7,582.18**	**57,034.82**	**-49,452.64**
Overall Total	**282.10**	**0.00**	**282.10**

Caretaker Farm Proposed Budget (1/1/97–12/31/97)

Category	Budget
INCOME	
FARM SHARES	
Half-shares	12025.00
Shares	62500.00
Total income	$74525.00

EXPENSES	
ADMINISTRATION	
Advertising	105.00
Farm dues	250.00
Liability insurance	390.00
Office supplies	340.00
Postage	205.00
Printing	260.00
Land taxes	745.00
Telephone	215.00
Total administration	$2510.00
APPRENTICES:	
4 Apprentices	8640.00
FICA (7.65%)	661.00
Room & board	6720.00
Worker's compensation	900.00
Total apprentices	$16921.00
FARMERS:	
FICA (7.65%)	2148.00
Health insurance	3364.00
Pension (10%)	2808.00
Salary:	
Elizabeth	14040.00
Sam	14040.00
Total farmers	$36400.00

Caretaker Farm Proposed Budget (1/1/97–12/31/97)

OPERATING EXPENSES
Depreciation:

Building		3450.00
Drains		585.00
Machinery		3638.00
	Total depreciation	$7673.00

Diesel		290.00
Electricity		550.00
Green manure		980.00
Mulch straw		669.00
Propane		600.00
Repairs		1300.00
Seeds & plants		1250.00
Soil amendments		1900.00
Supplies		2620.00
	Total operating expenses	$17832.00

RESERVE	$862.00

Total expenses $74525.00

Brookfield Farm Budget Estimations FY 1997 (1/1/97–12/31/97)

EXPENSES

FARMER

Dan Kaplan salary		30000
FICA (7.65%)		2295
Health insurance		4000
Workers' compensation (6.9%)		2070
Pension fund (2%)		600
	Total farmer	**$38965**

APPRENTICES

3 full season (8 months@$600/month)		14400
0 summer season (3 months@$400/month)		0
Housing (12 months@$650/month)		7800
FICA (7.65%)		1102
Workers' compensation (6.9%)		994
Health insurance (27 months@$100/month)		2700
Lunches (480 meals@$2/meal)		960
Recruiting (listings)		25
Training (NOFA conference)		200
	Total apprentice program	**$28180**

LABOR

Hand weeding (80 days@$50/day)		4000
Bookkeeper (plus full share barter)		600
	Total casual labor	**$4600**

GENERAL

Maintenance/repair		4000
Fuel		2100
Supplies		3000
Vet		200
	Total general expenses	**$9300**

VEGETABLES

Green manure		250
Seeds and plants		3000
Greenhouse		1800
Field rental		400
Fertilizer/manure		1750
	Total vegetable production	**$7200**

PIGS

Piglets (5 pigs@$35/pig)		175
Feed (1500 lbs@$0.18/lb)		270
Processing		550
	Total pigs	**$995**

CATTLE

Hay (1100 bales@$1.00/bale)		1100
Slaughter and processing		300
Pasture and barn rental		550
	Total cattle	**$1950**

CHICKENS

Chicks (150 chicks@$1.00/chick)		150
Feed (1500 lbs@$0.18/lb)		270
Processing		275
	Total chickens	**$695**

SHEEP

Lambs (10 lambs@$35/lamb)		350
Feed		0
Processing and slaughter		250
	Total sheep	**$600**

Brookfield Farm Budget Estimations FY 1997 (1/1/97–12/31/97)

UTILITIES		
Electric		1200
Telephone		550
Water		300
Propane		900
Greenhouse telephone		250
Trash		150
Sani can		600
	Total utilities	**$3950**
ADMINISTRATION		
Taxes		300
Legal		400
Accounting		800
Insurance		650
Vehicle registration and insurance (van & truck)		1000
Office supplies		200
Postage		900
Copies (newsletter & mailings)		800
Dues and subscriptions		350
	Total administration	**$5400**
MARKETING		
Promotional items		0
Advertising (banner, miscellaneous)		200
Printing (brochure, stickers)		300
	Total marketing	**$500**
CONTINGENCY FUND		1500
DEBT REPAYMENT		3500
TOTAL REGULAR EXPENSES		**$107335**
CAPITAL EXPENSES		
Barn, mf 35, irrigation pipe		43400
	Total capital expenses	**$43400**
TOTAL EXPENSES		**$150735**

INCOME

SHARES		
Full shares (70@$480/full share)		33600
Half-shares (240@$280/half-share)		67200
	Total farm shares	**$100800**
BULK PRODUCE		500
SHOP SALES (bread, milk,apples,&c)		500
LIVESTOCK (2 calves@$200/calf)		400
MEAT		
Beef (500 lbs@$3/lb)		1500
Pork (500lbs@$3/lb)		1500
Chicken (550 lbs@$1.75/lb)		963
Sheep (400 lbs@$3/lb)		1200
	Total meat sales	**$5163**
PROMOTIONAL ITEMS (T-shirts, mugs, &c)		100
TOTAL INCOME		**$107463**

Appendix D

Sample Prospecti

40 Acres & Ewe
339 Tenth Street
Prairie Farm, WI 54762
(715) 455-1663

40 Acres & Ewe
A Community Supported Farm

T he act of [eating] what the earth has grown is perhaps your most direct interaction with the earth.

—*Frances Moore Lappé*

What Is CSA?

Community Supported Agriculture (CSA) is a way for consumers and farmers to share the risks and benefits of sustainable agriculture. In its simplest form, **Community Supported Agriculture** is an agreement between one or more farms and a group of consumer members. Each growing season (June – October) members pay a pre-determined price up front to support the farm. In return, the members receive an agreed-upon share of the farm's output.

Community Supported Agriculture offers new opportunities to provide predictable income to small-scale family farms. Several factors help ensure that CSA farmers are able to utilize sustainable farming practices:

- Share price reflects the cost of environmentally sound production.
- Receiving payment up front, the farmers avoid the extra cost of borrowing operating capital.
- If weather or other factors result in more or less than the expected output, members share equally in the losses or abundance.

CSAs emphasize the role of the consumer in consciously taking moral responsibly for the care of the land, animals, and people that produce the food human beings need. **Community Supported Agriculture** has been stimulated by consumer interest in locally produced organic food, as well as environmentally and socially conscious values that recognize the contribution of healthy farming activity to *both* rural and urban communities.

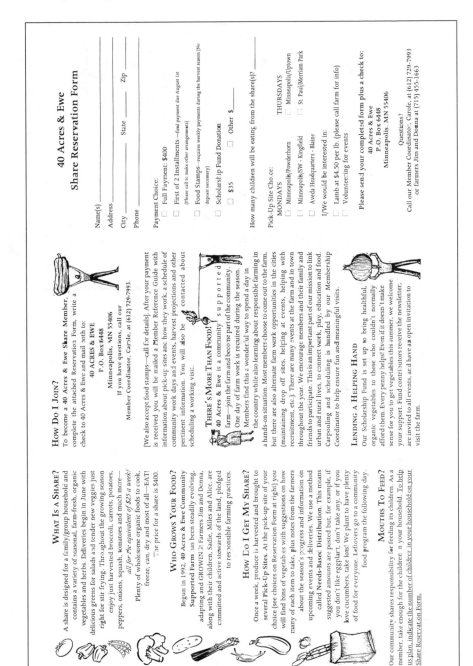

WHAT IS A SHARE?

A share is designed for a family/group household and contains a variety of seasonal, farm-fresh, organic vegetables and herbs. Deliveries begin in June with delicious greens for salads and tender new veggies just right for stir frying. Throughout the growing season enjoy just harvested broccoli, carrots, potatoes, peppers, onions, squash, tomatoes and much more—*all for the equivalent of $25 a week!* Plenty of wholesome organic foods to cook, freeze, can, dry and most of all—EAT! The price for a share is $400.

WHO GROWS YOUR FOOD?

Begun in 1992, **40 Acres & Ewe Community Supported Farm** has been steadily evolving, adapting and GROWING! Farmers Jim and Donna, along with their children, Sadie, Miles and Alice, are committed and active stewards of the land, pledged to responsible farming practices.

HOW DO I GET MY SHARE?

Once a week, produce is harvested and brought to several Pick-Up Sites. At the pick-up site of your choice (see choices on Reservation Form at right) you will find bins of vegetables with suggestions on how many of each item to take, plus notes from the farmers about the season's progress and information on upcoming events and deliveries. We use a method called **Needs-Based Distribution**. This means suggested amounts are posted but, for example, if you don't like eggplant, don't take any, or if you love cucumbers, take lots! We plant to have plenty of food for everyone. Leftovers go to a community food program the following day.

MOUTHS TO FEED?

Our community shares responsibility for feeding its children. As a member, take enough for the children in your household. To help us plan, indicate the number of children in your household on your Share Reservation Form.

HOW DO I JOIN?

To become a 40 Acres & Ewe Share Member, complete the attached Reservation Form, write a check to 40 Acres & Ewe and mail both to:

40 ACRES & EWE
P.O. Box 6448
Minneapolis, MN 55406
If you have questions, call our
Member Coordinator, Carole, at (612) 729-7993.

[We also accept food stamps—call for details]. After your payment is received you will be mailed a Member Reference Guide with information about pick-up sites and how they work, a schedule of community work days and events, harvest projections and other pertinent information. You will also be contacted about scheduling a working visit.

THERE'S MORE THAN FOOD!

40 Acres & Ewe is a community s u p p o r t e d farm—join the farm and become part of the community. One day of farm work is required during the season. Members find this a wonderful way to spend a day in the country while also learning about responsible farming in a hands-on situation. Most members choose to come out to the farm, but there are also alternate farm work opportunities in the cities (maintaining drop off sites, helping at events, helping with recruitment, etc.). There are many events at the farm and in town throughout the year. We encourage members and their family and friends to participate. This is an important part of our mission to link urban and rural lives, to connect work, play, education and food. Carpooling and scheduling is handled by our Membership Coordinator to help ensure fun and meaningful visits.

LENDING A HELPING HAND

Our Scholarship Fund is set up to bring healthful, organic vegetables to those who couldn't normally afford them. Every penny helps! Even if it doesn't make sense for you to get vegetables this summer, we welcome your support. Fund contributors receive the newsletter, are invited to all events, and have an open invitation to visit the farm.

40 Acres & Ewe
Share Reservation Form

Name(s) _____

Address _____

City _____ State _____ Zip _____

Phone _____

Payment Choice:

☐ Full Payment: $400

☐ First of 2 Installments —final payment due August 1st
(Please call to make other arrangements)

☐ Food Stamps - requires weekly payments during the harvest season [No deposit necessary]

☐ Scholarship Fund Donation

☐ $35 ☐ Other $ _____

How many children will be eating from the share(s)? _____

Pick-Up Site Choice:

MONDAYS

☐ Minneapolis/Powderhorn

☐ Minneapolis/SW - Kingfield

☐ Aveda Headquarters - Blaine

THURSDAYS

☐ Minneapolis/Uptown

☐ St. Paul/Merriam Park

I/We would be interested in:

☐ Lamb at $4.50 per lb. (please call farm for info)

☐ Volunteering for events

Please send your completed form plus a check to:

40 Acres & Ewe
P.O. Box 6448
Minneapolis, MN 55406

Questions?
Call our Member Coordinator, Carole, at (612) 729-7993
or farmers Jim and Donna at (715) 455-1663

40 Acres and Ewe Prospectus

The Agreement:
I would like to be a shareholder in the *Angelic Organics* Community Supported Agriculture Project. I understand that the farm workers will do their best to provide all they have promised, and I agree to excuse them for the curveball acts of God that might trip them up.

(Signature) _____ (Date) _____

Necessary Info:
(Name or Names–*Please Print*)

(Address)

(City) _____ (State) _____ (Zip) _____

(Telephone–Home)

(Telephone–Fax) _____ (E-Mail Address)

If you were a 1996 Shareholder, what was your:
(1996 Box Name?) _____ (1996 Delivery Site?)

Please choose your 1997 Box and Delivery Information (You'll recognize your weekly box at your delivery site, because it will have the name that you chose on it. Projected delivery site locations are listed on the reverse side of this form)

(1997 Desired Box Name–no more than 10 letters) _____ (1997 Desired Delivery Site see reverse side for locations)

Shared Box Info: Please fill out the following if you will be sharing the box with another household and wish for this secondary household to participate in all of the benefits of being a CSA member (e.g., an extra copy of the newsletter (a $30 value), participation in our CSA services directory, being on our mailing list...). We suggest an **extra $20** to support our additional costs that go along with sharing a box.

(Name or Names–*Please Print*)

(Address)

(City) _____ (State) _____ (Zip) _____

(Telephone–Home) _____ (Telephone–Work)

(Telephone–Fax) _____ (E-Mail Address)

Vegetable Survey–please see the reverse side for details. Please place an "*M*" next to your 3 most desired and an "*L*" next to your 3 least desired vegetables. Due to computer and box label constraints, only the first 3 selections of each type will be recorded.

Cucumbers	Basil	Spinach
Eggplant	Cilantro	Celery and Celeriac
Melons	Parsley	Leeks
Peppers	All Herbs	Onions	Zucchini & Summer Squash
Sweet Corn	Broccoli	Beets	Salad Mix (Lettuce and
Tomatoes	Cabbage	Car...	Baby Greens)
Winter Squash	Rad..shes	Pota...	Cooking Greens (Chard, Kale, Chinese Cabbage...)

1997 SIGN-UP FORM
Angelic Organics
8/96

Pricing Info:
Basic Box Price for 20 Weeks of Fresh, Organic, Biodynamic Vegetables

If you pay between....	The price is....
8/1/96 - 8/31/96	$380
9/1/96 - 9/30/96	$390
10/1/96 - 11/15/96	$400
after 11/15/96	$420

Gladiola Share
$80 for 10-12 weeks,
6-8 glads per week

Alternate Payment Plans:
♦ If you wish to divide the payment, please write out 2 checks, each covering half of the Total Payment amount. Date one check for today; post date the 2nd check for a date that is convenient for you up until 3 months from the date of your 1st check. **Send both checks with this sign-up form.** We will not deposit the 2nd check until its date. To receive a discounted price, your 1st check must be dated within a time period specified above.
♦ If the above payment plan still doesn't work for you, but you simply must have a share, call us. We can probably work something out.

Payment Info:
Please make your check payable to:
♦ Angelic Organics

Please send check(s) and sign-up form to:
♦ Angelic Organics,
1547 Rockton Rd.,
Caledonia, IL. 61011

or sign up through our web page:
http://www.Angelic-Organics.com
or fax your sign-up form to:
815-389-3106
(your check must be received within 1 week of your sign-up in order to reserve your space)

Angelic Caring:
If you desire, please complete this section and fill in the amount to the right. ♦ ♦ ♦ ♦
[ ] We want to help make sure that Angelic Organics is growing food for us in the years to come.

Levels of Support:
$15–seed $100–vine
$25–cotyledon $500–pumpkin blossom
$50–seedling $1000–giant pumpkin
$other–please detail _____

Last Bit of Info:
[ ] I have broken my payment into more than one check and have mailed both the current and post dated check as in the **Payment Info** instructions above.

Based on the items described on the reverse side of this page, my responses follow:
[ ] I would consider hosting a delivery site as described on the reverse side. (If we choose your home or business as a delivery site, we'll offer you either a free gladiola share or $70 off your box.)
[ ] I will volunteer for the following area(s): [ ]Field [ ]Cook [ ]Referral [ ]Delivery
[ ] Please raise beans for us. We plan to harvest them [ ] 1 time [ ] 2 times
[ ] I choose to not receive conventionally grown (non-organic–raised on John's sister's farm) sweet corn in my box. Please substitute yummy Angelic Organic vegetables in place of the corn.

Any Questions? Contact the Farm: ♦ Telephone: 815-389-2746 ♦ Fax: 815-389-3106 ♦ Web Site: http://www.Angelic-Organics.com ♦ E-Mail: AngelicOR@aol.com

Figuring the Cost:
Base Price
($380, $390, $400, $420)
–see pricing info above) $_____

Add: Glad Share Cost
(If you want a glad share, add $80 for each share–see pricing info above) $_____

Add: Shared Box Contribution
(We suggest, but do not require, adding $20 if you are sharing a box –see Shared Box Info section) $_____

Add: Angelic Care
(see Angelic Caring Section on the immediate left for details) $_____

Add: Multiple Year Boxes
(You may elect to purchase boxes for future years and lock into the price of $380 per year–see Multiple Box Info Section on the reverse) $_____

Subtract: Volunteer Perk
(you are entitled to take $20 off the base price for each full day you helped Angelic Organics during 1996) $_____

Total Payment $_____

Multiple Year Sign Ups: We are offering a long term subscription (more than one year) at a locked in price of $380 per year. Not only are you protected from future price increases, but you are guaranteed a membership in our increasingly popular CSA. The proceeds from these multi-year subscriptions will finance long term capital projects, such as a new irrigation system, or new barn roofs. Please enter the appropriate amount on the *Multiple Year Boxes* line in the **Figuring the Cost** section on the reverse side of this page. For example, an appropriate amount to sign up for a share for 1997, 1998, 1999, 2000, & 2001 would be $1,900.

Volunteering: Angelic Organics is a Community Supported Agriculture (CSA) farm. Part of the community support can come in the form of volunteer labor. Although shareholders are not required to work for the farm, their assistance is especially meaningful. If you are interested in volunteering in any of the following capacities, please indicate by checking the appropriate box on the reverse side of this page (under **Last Bit of Info**).

♦ **Field Work**—help us out in the field
♦ **Shareholder Referral**—share your CSA experience with potential new members
♦ **Delivery Assistance**—help our driver deliver boxes on a Wednesday or a Saturday
♦ **Guest Cook**—bring your culinary skills into the kitchen & feed the crew for a day
♦ **Clerical Work**—help with a mailing, filing or some other routine office task
♦ **CSC**—Angelic Organics Community Support Committees—formed by our shareholders to help ensure Angelic Organics sustainability.

Vegetable Survey. We want to know which crops shareholders are the most and least excited about. This will allow us to plan our production accordingly. There will be certain times, however, when we will give a surplus vegetable to those who have indicated their favorite. Please place an "M" next to your 3 most favorites and an "L" next to your 3 least favorites. Due to computer and box label constraints, only the first 3 selections of each type will be recorded.

Green Beans: due to labor constraints, we will only raise green beans for shareholders who harvest them themselves. Beans are usually available from mid-July through mid-August. Please indicate your desire to participate in the bean harvest on the reverse side of this page (under **Last Bit of Info**)

Sweet Corn: the sweet corn that we put in your box may not be raised at Angelic Organics It may be raised by Farmer John's mother and sister using conventional methods (chemicals). The distinction of this possibly non-organic sweet corn is that it is very fresh when you receive it. It is the only vegetable we include in your box that is not raised according to organic guidelines. If you would **not** like to receive conventionally grown sweet corn in your box and would prefer to substitute yummy Angelic Organic vegetables in their place, please indicate so on the other side of this page (under **Last Bit of Info**). If we are able to grow organic sweet corn, we will indicate so in the newsletter and provide it to all shareholders.

Credit Cards: due to the added expense for processing credit card payments (to the banks), we do not currently accept credit cards as payment for shares.

Macrobiotic Customization: due to the low interest in our Macrobiotic option (which excluded nightshades–tomatoes, peppers, eggplants, & potatoes), we have **discontinued** offering the macrobiotic option as described in our brochure.

Any Questions? Contact the Farm:	
♦ Telephone: 815-389-2746	♦ Web Site: http//www.Angelic-Organics.com
♦ Fax: 815-389-3106	♦ E-Mail: AngelicOR@Angelic-Organics.com

Delivery Sites: The following list identifies where we delivered boxes for the 1996 season. Most delivery sites are porches or garages at a shareholder's home. The sites are subject to change in 1997. The receipt of your sign up will be confirmed to you within 2 months after your payment is received. Your exact site as well as your sign-up options will be detailed and confirmed to you approximately one month before we start to deliver your vegetables (confirmation date: approximately 5/25/96; 1st delivery: approximately June 25, 1997). Please indicate where you would like your box delivered on the reverse side of this sheet. From the scheduled delivery time, you are given 24 hours to pick up your box. Most sites do not allow you to pick up your box between 10:00 pm and 7:00 am.

Site/Neighborhood — Approximate Location

SATURDAY DELIVERIES–all sites in the Chicago city limits.

♦ **Bucktown (BT)** near 1756 W. Cortland, Chicago, IL 60622
♦ **South Loop (SL)** 637 S. Dearborn, Chicago, IL 60605 (Grace Place)
♦ **Hyde Park (HY)** near 5400 S. Kenwood, Chicago IL 60615
♦ **Lincoln Park (LP)** near 1100 W. Wellington, Chicago IL 60657
♦ **Wrigleyville (WV)** 1346 W. Waveland, Chicago, IL 60613 (Waveland Wellness)
♦ **Ravenswood Manor (RM)** near 2800 W. Leland, Chicago IL 60625
♦ **Edgewater (EW)** near 1500 W. Victoria, Chicago, IL 60660
♦ **Rogers Park–West (WR)** near 6400 N. Albany, Chicago IL 60645

WEDNESDAY DELIVERIES–Chicago Suburbs & a new Wednesday Chicago Delivery

♦ **Schaumburg (SC)** 15 E. Scully, Schmbrg, IL 60193 (Healthmasters just off Roselle Road)
♦ **Highland Park (HP)** near 1600 Cloverdale, Highland Park, IL 60035
♦ **Evanston–North (EN)** near 2200 Crawford, Evanston, IL 60201
♦ **Evanston–South (ES)** near 1200 Washington, Evanston, IL 60202
♦ **Chicago–Wednesdays(CW)** near 6600 N. Talman, Chicago, IL 60645
♦ **Park Ridge (PR)** near 300 N. Ashland Ave, Park Ridge IL 60068
♦ **Oak Park (OP)** near 400 N. Harvey, Oak Park IL 60302
♦ **Villa Park (VP)** 1555 S. Ardmore, V.P., 60181 (Good Nature 1/4 block from Roosevelt Rd)
♦ **Geneva (GV)** near 100 Ford St., Geneva IL 60123

WEDNESDAY DELIVERIES–North Central Illinois

♦ **Beloit (BE)** near 800 Wisconsin, Beloit, WI 53511
♦ **Roscoe (RC)** 5647 Elevator Rd, Roscoe, IL 61073 (Weigh & Pay Farm Stand)
♦ **Rockford–North East(RFE)** near 3000 Burfwod Dr, Rkfrd, IL 61114 (Benbury Subdivision)
♦ **Rockford–North Central (RFC)** near 4000 Coventry Drive, Rockford, IL 61114
♦ **Belvidere** contact the farm for details

Hosting or Adding a Delivery Site: There is always the possibility that a site may not be available from year to year. Or there may be a new site which better suits the demographics of the shareholding population. In any case, we are always interested in knowing whether there are any sites which our shareholders may be willing to offer for delivery usage (allowing for pickup up to 24 hours after delivery with no pickups between 10:00 pm and 7:00 am). If you are interested in being a drop site host and receiving either a $70 share discount or a free Gladiola share, please check the box on the reverse side under "Last Bit of Info". We will contact you to discuss the logistics and kind necessary to be used as a site. Also, if you are interested in establishing a new site, in a town other than listed above, and have about 15 people interested with 10 definite boxes, please contact the farm at 815-389-2746 to ensure that: a) your area is somewhat near our established delivery route, and b) we still have the necessary number of shares available for new shareholders.

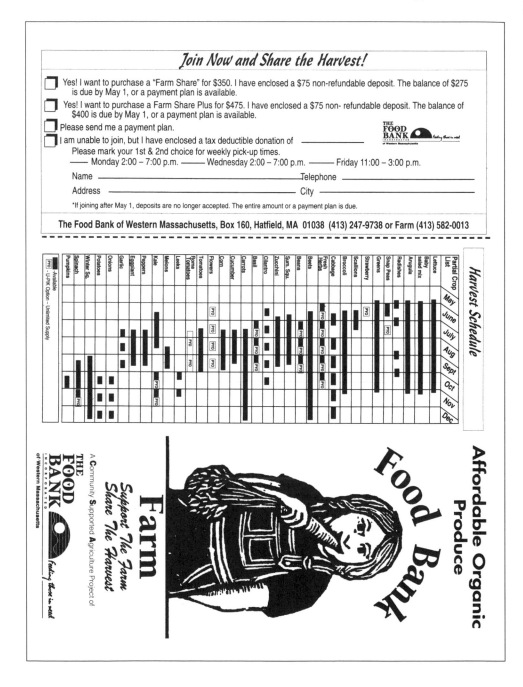

Join Now and Share the Harvest!

☐ Yes! I want to purchase a "Farm Share" for $350. I have enclosed a $75 non-refundable deposit. The balance of $275 is due by May 1, or a payment plan is available.

☐ Yes! I want to purchase a Farm Share Plus for $475. I have enclosed a $75 non- refundable deposit. The balance of $400 is due by May 1, or a payment plan is available.

☐ Please send me a payment plan.

☐ I am unable to join, but I have enclosed a tax deductible donation of ———————

Please mark your 1st & 2nd choice for weekly pick-up times.

—— Monday 2:00 – 7:00 p.m. —— Wednesday 2:00 – 7:00 p.m. —— Friday 11:00 – 3:00 p.m.

Name ————————————————————— Telephone ————————————

Address ————————————————————— City ————————————

*If joining after May 1, deposits are no longer accepted. The entire amount or a payment plan is due.

The Food Bank of Western Massachusetts, Box 160, Hatfield, MA 01038 (413) 247-9738 or Farm (413) 582-0013

THE FOOD BANK INCORPORATED of Western Massachusetts

Harvest Schedule

Affordable Organic Produce

Food Bank Farm

A Community Supported Agriculture Project of

THE FOOD BANK INCORPORATED of Western Massachusetts

Support The Farm Share The Harvest

Food Bank Farm CSA Prospectus (edited for publication)

Become Part of The Farm

Since 1992, the Food Bank has provided hundreds of satisfied shareholders with a steady supply of premium-quality organic produce. At the same time, the farm has also grown over 270,000 lbs. of food for those that are hungry in our community. When you buy a share in the Farm you will enjoy the freshest organic produce at a reasonable cost, support local agriculture and help to feed those in need.

Reap a Weekly Harvest

From late May through November, you can pick up your vegetables at the Farm on Monday or Wednesday between 2:00-7:00 or on Friday from 11:00-3:00. You will choose from over 50 varieties of vegetables, fruit, herbs and flowers. In November and December, a month's supply of winter crops (carrots, potatoes, and more) will be distributed from our root cellar. Our diversity of crops, and extensive irrigation system insure against crop failure.

Share the Satisfaction . . .

The Food Bank Farm is about building community. On your weekly pick-up you will see where your food comes from and meet the folks who grow it. You will experience the harmony of the seasons. Sugar snap peas will announce the arrival of spring, followed by ripe tomatoes and sweet corn in the summer, and butternut squash in the fall. You will also be joining in our effort to feed those less fortunate.

Your share of the harvest provides nutritious foods for:

- the unemployed family at a Springfield soup kitchen
- the battered wife in a shelter in Greenfield
- the mother feeding her family from the Northampton Survival Center
- the hilltown widow trying to get by on a social security check

Organic Produce for You !

The cost of a FarmShare is $350(approx. 10-20 lbs. @ $12/wk). A Farm Share Plus is also available for $475. The FarmShare will feed a family of 3-5. The Farm Share Plus will feed 5-7 people. Shares will be smaller in the spring and will increase and expand in variety as the season progresses.

There is no work requirement. Produce will be picked and washed for you. Certain crops are available for unlimited U-Pic. Volunteers are welcome.

Organic and Affordable !

The FOOD BANK FARM is an inexpensive way to ensure the safety of your food supply. See for yourself these findings based on a "1995" independent study.*

• Share Plus at the Farm	$450.00
• Supermarket - Nonorganic	$715.93
• Natural Foods Supermarket - Organic	$1021.38

* Comparison based on comparable produce amounts. Price data compiled on a weekly basis.

What is a CSA?

Community Supported Agriculture farms (CSAs) are supported by members who buy shares in a farm and in return receive fresh produce throughout the season. Originating in Europe, there are now over 500 CSAs in the United States. The Food Bank Farm is a unique model for other communities. It is one of the largest and most efficient CSAs in the country and the first to help feed the hungry in our communities.

Where is It Located?

The 60 acre Farm is located at 115 Bay Road (Rt.47) in Hadley, just 5 minutes from Northampton and 10 minutes from Amherst. Visitors are always welcome.

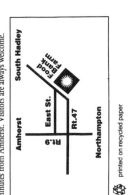

♻ printed on recycled paper

Food Bank Farm CSA Prospectus (edited for publication)

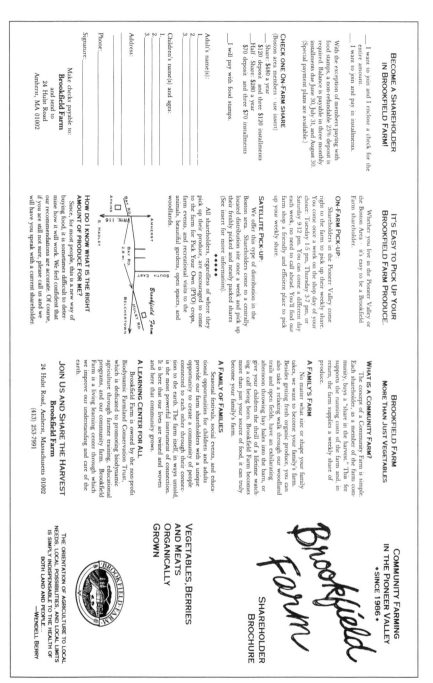

BECOME A SHAREHOLDER IN BROOKFIELD FARM!

___ I want to join and I enclose a check for the entire amount.

___ I want to join and pay in installments.

With the exception of members paying with food stamps, a non-refundable 25% deposit is required. Balance is payable in three monthly installments due June 30, July 31, and August 30. (Special payment plans are available.)

___ I will pay with food stamps.

CHECK ONE ON-FARM SHARE
(Boston area members - use insert)

___ Share: $480 a year
$120 deposit and three $120 installments

___ Half Share: $280 a year
$70 deposit and three $70 installments

Adult's name(s):
1. _____
2. _____
3. _____

Children's name(s) and ages:
1. _____
2. _____
3. _____

Address: _____

Phone: _____

Signature: _____

Make checks payable to:
Brookfield Farm
and send to
24 Hulst Road
Amherst, MA 01002

IT'S EASY TO PICK UP YOUR BROOKFIELD FARM PRODUCE.

Whether you live in the Pioneer Valley or the Boston Area · it's easy to be a Brookfield Farm shareholder.

ON-FARM PICK-UP:

Shareholders in the Pioneer Valley come right to the farm to pick up their weekly share. You come once a week on the shop day of your choice: Tuesday 1-5 pm, Thursday 3-7 pm, or Saturday 9-12 pm. You can come a different day each week, no need to call ahead. You'll find our farm shop a friendly and efficient place to pick up your weekly share.

SATELLITE PICK-UP:

We offer this type of distribution in the Boston area. Shareholders come to a centrally located distribution site once a week and pick up their freshly picked and neatly packed shares (See insert for more information).

✦✦✦✦✦

All shareholders, regardless of where they pick up their produce, are encouraged to come to the farm for Pick Your Own (PYO) crops, farm events, and recreational visits to the animals, beautiful gardens, open spaces, and woodlands.

HOW DO I KNOW WHAT IS THE RIGHT AMOUNT OF PRODUCE FOR ME?

Since, for most people, this is a new way of buying food, it is sometimes difficult to determine how it will work. We feel confident that our recommendations are accurate. Of course, if you are still not sure, please call us and we will have you speak with a current shareholder.

BROOKFIELD FARM
MORE THAN JUST VEGETABLES

WHAT IS A COMMUNITY FARM?

The concept of a Community Farm is simple. Each shareholder, as a member of the farm community, buys a "share in the harvest." This fee supports the running costs of the farm and, in return, the farm supplies a weekly share of produce.

A FAMILY'S FARM

No matter what size or shape your family takes, we want to become your family's farm. Besides getting fresh, organic produce, you can also take a relaxing walk through our woodland trails and open fields, have an exhilarating afternoon throwing hay bales into the barn, or give your children the thrill of a lifetime watching a calf being born. Brookfield Farm becomes more than just your source of food, it can truly become your family's farm.

A FAMILY OF FAMILIES

Seasonal festivals, social events, and educational opportunities for children and adults provide our farm shareholders with a unique opportunity to create a community of people connected to each other through their connection to the earth. The farm itself, in ways untold, is the most powerful instrument of connection. It is here that our lives are twined and woven and here that community grows.

A LEARNING CENTER FOR ALL.

Brookfield Farm is owned by the non-profit Biodynamic Farmland Conservation Trust, which is dedicated to promoting biodynamic agriculture through farmer training, educational programs, and our community farm. Brookfield Farm is a living learning center through which we improve our understanding and care of the earth.

JOIN US AND SHARE THE HARVEST
Brookfield Farm
24 Hulst Road, Amherst, Massachusetts 01002
(413) 253-7991

COMMUNITY FARMING IN THE PIONEER VALLEY
✦ SINCE 1986 ✦

Brookfield Farm

SHAREHOLDER BROCHURE

VEGETABLES, BERRIES AND MEATS ORGANICALLY GROWN

THE ORIENTATION OF AGRICULTURE TO LOCAL NEEDS, LOCAL POSSIBILITIES, AND LOCAL LIMITS IS SIMPLY INDISPENSABLE TO THE HEALTH OF BOTH LAND AND PEOPLE....
—WENDELL BERRY

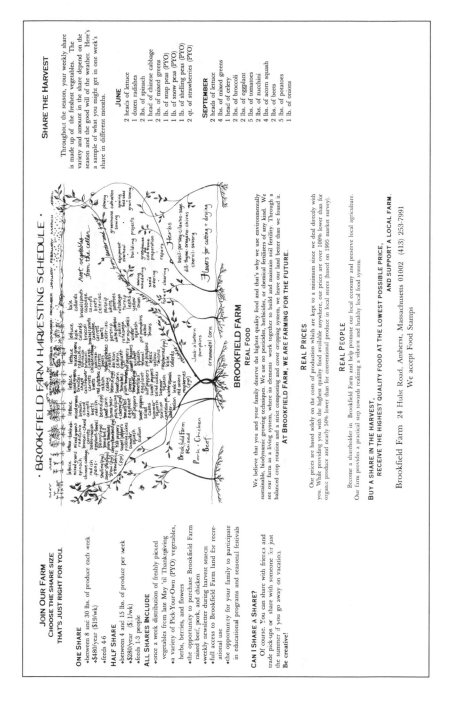

SHARE THE HARVEST

Throughout the season, your weekly share is made up of the freshest vegetables. The variety and amount in the share depend on the season and the good will of the weather. Here's a sample of what you might get in one week's share in different months.

JUNE
- 2 heads of lettuce
- 1 dozen radishes
- 2 lbs. of spinach
- 1 head of chinese cabbage
- 2 lbs. of mixed greens
- 1 lb. of snap peas (PYO)
- 1 lb. of snow peas (PYO)
- 1 lb. of shelling peas (PYO)
- 2 qt. of strawberries (PYO)

SEPTEMBER
- 2 heads of lettuce
- 4 lbs. of mixed greens
- 1 head of celery
- 2 lbs. of broccoli
- 2 lbs. of eggplant
- 5 lbs. of tomatoes
- 2 lbs. of zucchini
- 4 lbs. of acorn squash
- 2 lbs. of beets
- 5 lbs. of potatoes
- 1 lb. of onions

· BROOKFIELD FARM HARVESTING SCHEDULE ·

BROOKFIELD FARM

REAL FOOD

We believe that you and your family deserve the highest quality food and that's why we use environmentally sustainable, biodynamic growing techniques. We use no pesticides, herbicides, or chemical fertilizers of any kind. We see our farm as a living system, where its elements work together to build and maintain soil fertility. Through a balanced crop rotation and a strict composting and cover cropping system, we leave our land better than we found it. **AT BROOKFIELD FARM, WE ARE FARMING FOR THE FUTURE.**

REAL PRICES

Our prices are based solely on the costs of production which are kept to a minimum since we deal directly with you. While providing you with the highest quality food available anywhere, our prices are over 100% lower than for organic produce and nearly 50% lower than for conventional produce in local stores (based on 1995 market survey).

REAL PEOPLE

Become a shareholder in Brookfield Farm and help promote our local economy and preserve local agriculture. Our farm provides a practical step towards realizing a vibrant and healthy local food system.

BUY A SHARE IN THE HARVEST,
RECEIVE THE HIGHEST QUALITY FOOD AT THE LOWEST POSSIBLE PRICE,
AND SUPPORT A LOCAL FARM.

Brookfield Farm   24 Hulst Road, Amherst, Massachusetts 01002   (413) 253-7991
We accept Food Stamps

JOIN OUR FARM
CHOOSE THE SHARE SIZE
THAT'S JUST RIGHT FOR YOU!

ONE SHARE
- between 8 and 30 lbs. of produce each week
- $480/year ($19/wk)
- feeds 4-6

HALF SHARE
- between 4 and 15 lbs. of produce per week
- $280/year ($11/wk)
- feeds 1-3 people

ALL SHARES INCLUDE
- once a week distribution of freshly picked vegetables from late May 'til Thanksgiving
- a variety of Pick-Your-Own (PYO) vegetables, herbs, berries, and flowers
- the opportunity to purchase Brookfield Farm raised beef, pork, and chicken
- weekly newsletter during harvest season
- full access to Brookfield Farm land for recreational use
- the opportunity for your family to participate in educational programs and seasonal festivals

CAN I SHARE A SHARE?
Of course. You can share with friends and trade pick-ups or share with someone for just the summer if you go away on vacation. Be creative!

Brookfield Farm Prospectus (edited for publication)

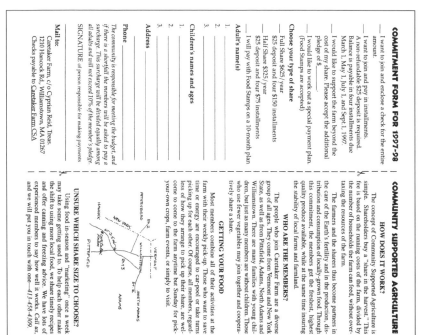

COMMITMENT FORM FOR 1997-98

___ I want to join and enclose a check for the entire amount.

___ I want to join and pay in installments.
A non-refundable $25 deposit is required.
Balance is payable in four installments due March 1, May 1, July 1, and Sept.1, 1997.

___ I would like to support the farm beyond the cost of my share. Please accept the additional pledge of $____.

___ I would like to work out a special payment plan. (Food Stamps are accepted)

Choose your type of share

___ Full Share $625/year
___ $25 deposit and four $150 installments
___ Half Share $325/year
___ $25 deposit and four $75 installments
___ I will pay with Food Stamps on a 10-month plan

Adult's name(s)

1.
2.
3.

Children's names and ages

1.
2.
3.

Address

Phone

SIGNATURE of person responsible for making payments

The community is responsible for meeting the budget, and if there is a shortfall the members will be asked to pay a surcharge. This surcharge will be divided equally among all adults and will not exceed 10% of the member's pledge.

Mail to:

Caretaker Farm, c/o Cyprian Reid, Treas.
1210 Hancock Rd., Williamstown, MA 01267
Checks payable to Caretaker Farm CSA

COMMUNITY SUPPORTED AGRICULTURE

HOW DOES IT WORK?

The concept of Community Supported Agriculture is simple. Shareholders buy a "share in the harvest." This fee is based on the running costs of the farm, divided by the number of households the farm can feed, without over-taxing the resources of the farm.

The farmers and the sharers thus become partners in the care of the Earth's fertility and in the production, distribution and consumption of locally-grown food. Through this commitment, the members get the freshest, highest quality produce available, while at the same time insuring the stability of local farms and farmland.

WHO ARE THE MEMBERS?

The people who join Caretaker Farm are a diverse group of all ages. They come from Vermont and New York State, as well as from Pittsfield, Adams, North Adams and Williamstown. There are many families with young children, but just as many members are without children. Those who use fewer vegetables may join together and cooperatively share a share.

GETTING YOUR FOOD

Most members combine all of their activities at the farm with their weekly pick-up. Those who want to save time or energy are encouraged to car-pool or take turns picking up for each other. Of course, all members, regardless of how they arrange to pick up their share, are welcome to come to the farm anytime but Sunday for pick-your-own crops, farm events, or simply to visit.

UNSURE WHICH SHARE SIZE TO CHOOSE?

Using food in-season and "marketing" once a week may take some getting used to. To help each other make the shift to using more local food, we share timely recipes and offer canning and freezing advice. We have lots of experienced members to say how well it works. *Call us,* and we will put you in touch with one—*413-458-4309.*

CARETAKER FARM

MORE THAN JUST VEGETABLES
A FARM OF TOMORROW

Established in 1970 by Sam and Elizabeth Smith, Caretaker Farm has practiced organic agriculture for over twenty-five years. Of its 35 acres, six are used for vegetable production, several acres of pasture support a small flock of sheep and an assortment of cows, pigs and chickens. The balance of the land is of woods, wetlands, ponds and stream. An orchard of apple, pear, and plum trees is being developed on a quarter acre. The farm also has its own bee hives and a bakery. In 1991, the farm became a community supported farm.

LAND

Food quality is directly connected to the health of the soil in which it is grown. We see our farm as a whole living system, where its elements work together to build and maintain soil fertility. Through a careful program of balanced crop rotation, composting and a cover-cropping system, we are farming for future generations.

COMMUNITY

Members not only share in the produce, but also the simple pleasures of participating in the seasonal life of the farm. On distribution days, in particular, the farm is alive with activity. Members greet each other and find out how things are growing from Sam, Elizabeth, or one of the farm's four seasonal apprentices. Children check on the well-being of their favorite animals. Happy pickers come up from the fields with baskets full of their pickings for the week. The farm itself, in ways untold, is a powerful instrument of connection.

CELEBRATION

Every day on the farm is cause for celebration, but seasonal festivities, social events, and educational opportunities are scheduled throughout the season. There is a great spirit of common purpose and fun at the Spring Open House where members plant potatoes for the year's crop. Come to Midsummer Night in July for an evening of music and dance. Gather together at the Harvest Festival to unearth the potatoes and pick pumpkins from the fields.

CARETAKER FARM
1210 Hancock Road
Williamstown, MA 01267
(413) 458-4309

CARETAKER FARM
Community Supported Agriculture

Shareholder
Brochure
1997

Caretaker Farm Prospectus—1997 (edited for publication)

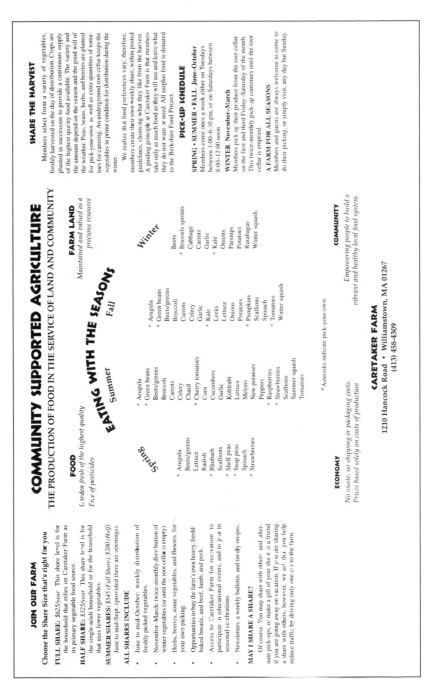

COMMUNITY SUPPORTED AGRICULTURE

THE PRODUCTION OF FOOD IN THE SERVICE OF LAND AND COMMUNITY

FOOD
Garden fresh of the highest quality
Free of pesticides

FARM LAND
Maintained and valued as a precious resource

EATING WITH THE SEASONS

Spring

* Arugula
   Beets/greens
   Lettuce
   Radish
   Rhubarb
   Scallions
* Shell peas
* Snap peas
   Spinach
* Strawberries

Summer

* Arugula
* Green beans
   Beets/greens
   Broccoli
   Carrots
   Celery
   Chard
* Cherry tomatoes
   Corn
   Cucumbers
   Garlic
   Kohlrabi
   Lettuce
   Melons
   New potatoes
   Peppers
* Raspberries
* Strawberries
   Scallions
   Summer squash
   Tomatoes

Fall

* Arugula
* Green beans
   Beets/greens
   Broccoli
   Carrots
   Celery
   Garlic
   Kale
   Leeks
   Lettuce
   Onions
* Pumpkins
   Potatoes
   Scallions
   Spinach
* Tomatoes
   Winter squash

Winter

   Beets
* Brussels sprouts
   Cabbage
   Carrots
   Garlic
* Kale
   Onions
   Parsnips
   Potatoes
   Rutabagas
   Winter squash

Asterisks indicate pick-own.

ECONOMY
No waste, no shipping or packaging costs.
Prices based solely on costs of production.

COMMUNITY
Empowering people to build a vibrant and healthy local food system.

CARETAKER FARM
1210 Hancock Road • Williamstown, MA 01267
(413) 458-4309

JOIN OUR FARM

Choose the Share Size that's right for you

FULL SHARE: *$625/year* This share level is for the household that relies on Caretaker Farm as its primary vegetable food source.

HALF SHARE: *$325/year* This share level is for the single-adult household or for the household that uses fewer vegetables.

SUMMER SHARES: *$545 (Full Share), $290 (Half)* June to mid-Sept. (provided there are openings).

ALL SHARES INCLUDE

* June to mid-October: weekly distribution of freshly picked vegetables.

* November–March: twice-monthly distribution of winter vegetables (or until the root cellar is empty).

* Herbs, berries, some vegetables, and flowers for your own picking.

* Opportunities to buy the farm's own honey, fresh baked breads, and beef, lamb, and pork.

* Access to Caretaker Farm for recreation to participate in educational events, and to join in seasonal celebrations.

* Newsletters, a weekly bulletin, and timely recipes.

MAY I SHARE A SHARE?

Of course. You may share with others and alternate pick-ups, or make a gift of your share to a friend if you are going away on vacation. If you are sharing a share with others, however, we ask that you help reduce traffic by driving only one car to the farm.

SHARE THE HARVEST

Members select from a variety of vegetables, freshly harvested on the day of distribution. Crops are planted in succession to provide a continuous supply of the highest quality food available. The variety and the amount depend on the season and the good will of the weather. Peas, beans, herbs, and berries are planted for pick-your-own as well as extra quantities of tomatoes for canning. An underground root cellar keeps the vegetables in prime condition for distribution during the winter.

We realize that food preferences vary; therefore, members create their own weekly share, within posted guidelines, choosing what they like from the harvest. A guiding principle at Caretaker Farm is that members take only as much food as they will use and leave what they do not want or need. All surplus food is donated to the Berkshire Food Project.

PICK-UP SCHEDULE

SPRING • SUMMER • FALL June–October
Members come once a week either on Tuesdays between 1:00–6:30 pm. or on Saturdays between 8:00–12:00 noon

WINTER November–March
Members pick up their produce from the root cellar on the first and third Friday–Saturday of the month. This twice-monthly pick-up continues until the root cellar is emptied.

A FARM FOR ALL SEASONS

Members and guests are always welcome to come to do their picking, or simply visit, any day but Sunday.

How You Can Receive a Good Humus Produce Harvest Box

Fill out the following form and send us a check for the your choice of share options. We will contact you to arrange pick up time at the location most convenient for you.

Name _____

Address _____

City _____

Phone _____

Share Options

☐ 1 month trial for $50*

☐ Quarterly for $150

☐ Half year for $300

☐ Full year for $600

*One time only.

Jeff & Anne Main
Good Humus Produce
12255 Road 84A
Capay, CA 95607

Community Supported Agriculture

with

a small organic
family farm

in

Capay, California

Good Humus Produce CSA Prospectus

WHAT IS COMMUNITY SUPPORTED AGRICULTURE?

CSA is an agreement between the members of the CSA and our farm to maintain a relationship for mutual benefit and support. As members you will receive a weekly supply of freshly picked organic produce throughout the year.

BENEFITS OF CSA

- Good tasting, unusual varieties of produce seldom found in standard stores.
- Guarantee of fresh, high quality, organic produce.
- Reasonable prices for your organically grown vegetables.
- One short stop for a variety of produce for the week.
- Assures you that your food dollar is having positive affect on ocal environmentally sound agriculture.
- Together we are creating an ecologically sustainable non-exploitative, regional system of agriculture.
- We would like to create an atmosphere of mutual trust anc interest in this common endeavor.
- Your participation & encouragement supports our ongoing stewardship of the land.

WHO IS GOOD HUMUS PRODUCE?

Jeff and Annie Main are fourth generation Californians who are continuing their agricultural heritage by caring for the earth in a respectful manner in order to preserve it for their three children and future generations. We have a 20 acre family farm located in the isolated Hungry Hollow Hills in Yolo County. We have been growing and marketing vegetables, fruits, herbs, and flowers organically in the Sacramento Valley since 1976. You can visit us weekly at the Saturday Davis Farmers Market or buy our products in the local co-ops in Davis and Sacramento.

WHAT WILL BE IN YOUR WEEKLY BOX?

The fruit, vegetables, and herbs you receive will depend on the season, but there will always be a good variety.

Summer crops may include apricots, tomatoes, squash, peaches, beans, melons, basil, garlic, onions, corn, cucumbers, nectarines, dried fruit and more.

Winter crops may include lettuce, salad mix, carrots, collards, kale, mustard greens, kohlrabi, fennel, broccoli, cauliflower, cabbage, winter squash, sun chokes, pumpkins and more.

FOR MORE
INFORMATION
CALL
(916) 787-3187

HARVEST BOX PICKUP

Weekly deliveries will be made in your area. Boxes not picked up will be donated to families in need. We will let you know of the time & locations for the box pickup. Provided you have at least 10 full shares we will deliver to your neighborhood.

PAYMENT

You can purchase shares of produce from Good Humus for:

1 month trial	$50*	4 deliveries
Quarterly	$150	12 deliveries
Half year	$300	25 deliveries
Full year	$600	50 deliveries

A $1 box deposit will be requested. Please return your boxes each week.

FARM VISITS

We would like to organize farm visits during the year for you to see our farm, work on any special projects, and to have a social meal together so we can meet each other. Please let us know if you are interested.

Good Humus Produce CSA Prospectus

Appendix E

Typical Shares From CSAs

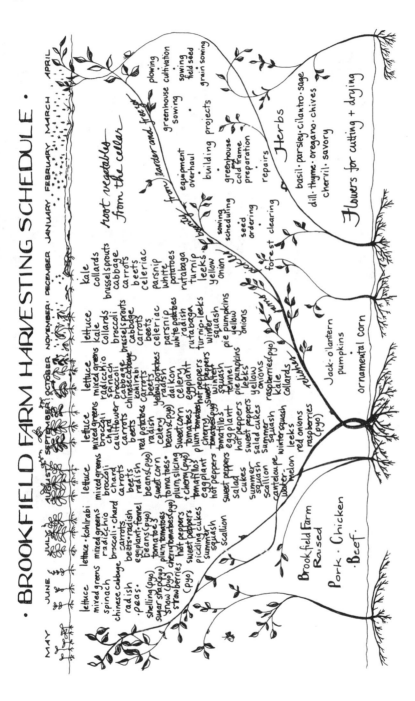

Brookfield Farm Harvesting Schedule—Typical Annual Planting

III. WHAT'S CELERIAC AND WHAT SHOULD I DO WITH ALL THESE VEGETABLES ANYWAY??

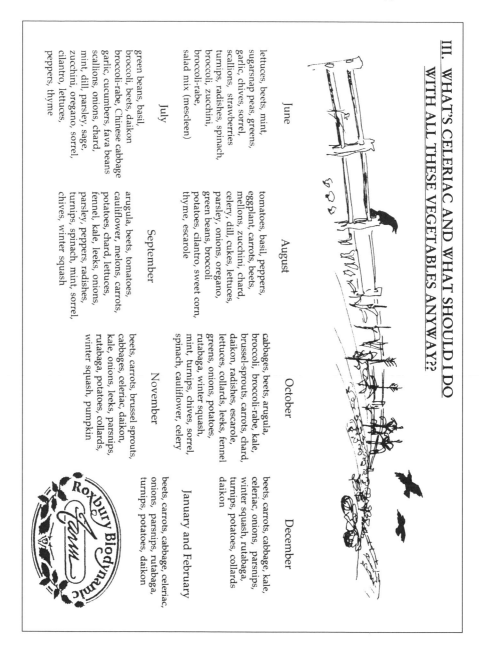

June
lettuces, beets, mint, sugarsnap peas, greens, garlic, chives, sorrel, scallions, strawberries turnips, radishes, spinach, broccoli, zucchini, broccoli-rabe, salad mix (mesclien)

July
green beans, basil, broccoli, beets, daikon broccoli-rabe, Chinese cabbage garlic, cucumbers, fava beans scallions, onions, chard, mint, dill, parsley, sage, zucchini, oregano, sorrel, cilantro, lettuces, peppers, thyme

August
tomatoes, basil, peppers, eggplant, carrots, beets, mellons, zucchini, chard, celery, dill, cukes, lettuces, parsley, onions, oregano, green beans, broccoli potatoes, cilantro, sweet corn, thyme, escarole

September
arugula, beets, tomatoes, cauliflower, melons, carrots, potatoes, chard, lettuces, fennel, kale, leeks, onions, parsley, peppers, radishes, turnips, spinach, mint, sorrel, chives, winter squash

October
cabbages, beets, arugula, broccoli, broccoli-rabe, kale, brussel-sprouts, carrots, chard, daikon, radishes, escarole, lettuces, collards, leeks, fennel greens, onions, potatoes, rutabaga, winter squash, mint, turnips, chives, sorrel, spinach, cauliflower, celery

November
beets, carrots, brussel sprouts, cabbages, celeriac, daikon, kale, onions, leeks, parsnips, rutabaga, potatoes, collards, winter squash, pumpkin

December
beets, carrots, cabbage, kale, celeriac, onions, parsnips, winter squash, rutabaga, turnips, potatoes, collards daikon

January and February
beets, carrots, cabbage, celeriac, onions, parsnips, rutabaga, turnips, potatoes, daikon

Roxbury Farm Produce Calendar—1996-97

Harvest Schedule

Partial Crop List	May	June	July	Aug	Sept	Oct	Nov	Dec
Lettuce	■	■	■	■	■	■		
Baby salad mix	■	■	■	■	■	■		
Arugula	■	■	■	■	■	■		
Radishes	■	■		■	■			
Snap Peas	■		PYO					
Greens	■	■	■	■	■	■		
Strawberry	PYO							
Scallions		■						
Broccoli		■	■	■	■	■		
Cabbage		■		■	■	■	■	■
Fresh Herbs		PYO	PYO	PYO	PYO	PYO		
Beets	■	■	■	■	■	■		
Beans			PYO	PYO	PYO			
Sum. Squ.			■	■	■			
Zucchini			■	■	■			
Cilantro		■	■	■	■	■		
Basil			PYO	PYO	PYO			
Carrots			■	■	■	■	■	
Cucumber			■	■				
Corn			■	■				
Flowers	PYO	PYO	PYO	PYO				
Tomatoes								
Roma Tomatoes				PYO	PYO			
Leeks						■	■	
Melons				■				
Kale		■	■			PYO	PYO	
Peppers			■	■				
Eggplant			■	■				
Garlic		■	■	■				
Onions						■	■	■
Potatoes						■	■	■
Winter Sq.					■	■	■	■
Spinach						■	PYO	
Pumpkins						■		

■ -Available
PYO - U-PIK Option – Unlimited Supply

Food Bank Farm Harvest Schedule

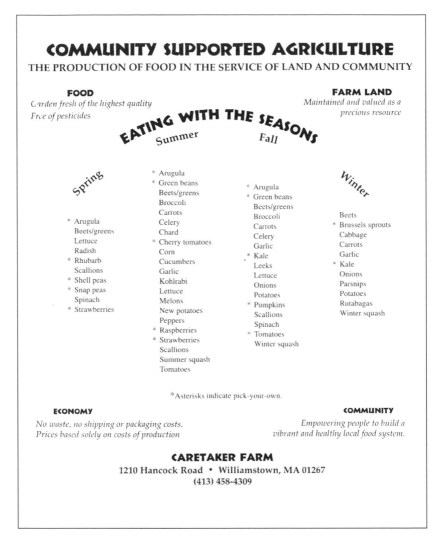

COMMUNITY SUPPORTED AGRICULTURE
THE PRODUCTION OF FOOD IN THE SERVICE OF LAND AND COMMUNITY

FOOD
Garden fresh of the highest quality
Free of pesticides

FARM LAND
Maintained and valued as a
precious resource

EATING WITH THE SEASONS
Summer   Fall

Spring

Winter

Summer
* Arugula
* Green beans
  Beets/greens
  Broccoli
  Carrots
  Celery
  Chard
* Cherry tomatoes
  Corn
  Cucumbers
  Garlic
  Kohlrabi
  Lettuce
  Melons
  New potatoes
  Peppers
* Raspberries
* Strawberries
  Scallions
  Summer squash
  Tomatoes

Spring
* Arugula
  Beets/greens
  Lettuce
  Radish
* Rhubarb
  Scallions
* Shell peas
* Snap peas
  Spinach
* Strawberries

Fall
* Arugula
* Green beans
  Beets/greens
  Broccoli
  Carrots
  Celery
  Garlic
* Kale
  Leeks
  Lettuce
  Onions
  Potatoes
* Pumpkins
  Scallions
  Spinach
* Tomatoes
  Winter squash

Winter
  Beets
* Brussels sprouts
  Cabbage
  Carrots
  Garlic
* Kale
  Onions
  Parsnips
  Potatoes
  Rutabagas
  Winter squash

*Asterisks indicate pick-your-own.

ECONOMY
No waste, no shipping or packaging costs.
Prices based solely on costs of production

COMMUNITY
Empowering people to build a
vibrant and healthy local food system.

CARETAKER FARM
1210 Hancock Road • Williamstown, MA 01267
(413) 458-4309

Caretaker Farm Seasonal Produce Calendar

SHARE THE HARVEST

Throughout the season, your weekly share is made up of the freshest vegetables. The variety and amount in the share depend on the season and the good will of the weather. Here's a sample of what you might get in one week's share in different months.

JUNE
2 heads of lettuce
1 dozen radishes
2 lbs. of spinach
1 head of chinese cabbage
2 lbs. of mixed greens
1 lb. of snap peas (PYO)
1 lb. of snow peas (PYO)
1 lb. of shelling peas (PYO)
2 qt. of strawberries (PYO)

SEPTEMBER
2 heads of lettuce
4 lbs. of mixed greens
1 head of celery
2 lbs. of broccoli
2 lbs. of eggplant
5 lbs. of tomatoes
2 lbs. of zucchini
4 lbs. of acorn squash
2 lbs. of beets
5 lbs. of potatoes
1 lb. of onions

Brookfield Farm Typical Weekly Share (edited for publication)

Appendix F

Resources

CSA Support and Consultation

The Biodynamic Farming and Gardening Association is an organization that has taken a leading role in the promotion and support of Community Supported Agriculture, and which provides a range of support and services. The Biodynamic Association is a nonprofit, membership organization whose task is to advance the principles and practices of biodynamic (BD) agriculture. Toward this end, the Association publishes a bi-monthly magazine entitled *BIODYNAMICS*, publishes books, offers a biodynamic advisory service, supports training programs, sponsors conferences and lectures, and funds research projects.

Biodynamic Preparations are available through the Josephine Porter Institute for Applied Biodynamics, Inc. Biodynamic farms or gardens are certified by the Demeter Association, Inc. Both are listed later in this appendix.

As a support to the community composting program, the Biodynamic Association offers for sale compostable food scrap bags: the Food Cycler™. The bags can be employed for community composting projects at schools, neighborhoods, retirement homes, college programs, and other places where people regularly congregate as a community of common interests, and also share food regularly.

Biodynamic Farming and Gardening Association
PO Box 550
Kimberton, PA 19442
Phone: 800-516-7797 or 610-935-7797
Fax: 610-983-3196

Community Food Security Coalition—Much more than just a new way of understanding the nation's hunger and farming problems, community food security unites under a single conceptual framework a series of strategies that otherwise have been separate. These include community gardening, farmers' markets, buying clubs and food coops, joint ventures between community development corporations and supermarket chains, food policy councils, CSAs, and others. The CFS Coalition provides networking and support, publishes a newsletter, and provides a range of services, including assistance to groups interested in applying for financial support under the Community Food Security Act.

As part of the 1996 Farm Bill, the Community Food Security Act was signed into US law. The Act makes available a total of $2.5 million a year until 2002 in small-scale matching grants to community food projects. The term "community food project" means a community-based project that requires a one-time infusion of federal assistance to become self-sustaining and that is designed to: (1) meet the food needs of low-income people; (2) increase the self-reliance of communities in providing for their own food needs; and (3) promote comprehensive responses to local food, farm and nutrition issues. Applicants must be non-profit organizations, but may use all available resources for matching funds, including those of for-profit entities. Contact the CFS Coalition for more information.

Community Food Security Coalition
PO Box 209
Venice, CA 90294
Phone/fax: 310-822-5410
E-mail: asfisher@aol.com

Community Supported Agriculture of North America, Inc. (CSANA) is a not-for-profit, educational, networking and technical assistance organization for existing, aspiring, and potential CSAs. CSANA publishes a booklet entitled *Basic Formula To Create a CSA,* a video entitled *It's Not Just About Vegetables,* a quarterly newsletter, and other resources of interest. As part of a SARE Peer Education and Mentoring grant, CSANA also provides a hotline for information and troubleshooting.

CSANA
Indian Line Farm
57 Jugend Road
Great Barrington, MA 01230
Phone/fax: 413-528-4374
E-mail: csana@bcn.net

CSA West is a program sponsored by the Community Alliance with Family Farmers (CAFF) and allied with the University of California at Santa Cruz. The CSA West program was founded to expand the CSA movement by offering regional information to farmers, prospective and current CSA members, the media, students and interns, and related non-profit organizations. Through apprenticeship programs, the organization teaches aspiring farmers the skills necessary to start their own CSA projects. CSA West strives to expand the exchange of new ideas, tried and true practices, and helpful tips on running CSAs among new, old, and potential participants in the western US. CSA West offers a comprehensive directory of California CSAs, general information packets, and guidelines for getting started. They have a library of relevant materials, a network of experts available for free consultation, and a speakers bureau.

CSA West
c/o Community Alliance with Family Farmers
PO Box 363
Davis, CA 95617
Phone: 916-756-8518

CSA Works is an organization set up to help farmers find the tools and techniques needed to run an efficient CSA enterprise. A for-profit corporation, the farmers who are involved in it all earn their livelihood growing food on CSAs. They charge a small consultation fee to cover their costs, based on a sliding scale. CSA Works does consulting to help CSA growers, and potentital CSA growers, develop a business plan, perform budget and cost analysis, determine row spacing for crops, optimize harvest and post-harvest operations, identify equipment needs, select specialized varieties, sell shares, maintain membership satisfaction, streamline distribution systems, and other services. CSA Works has produced a model of a Scrip Card for CSAs—*"In Farms We Trust"*—on Microsoft Word 2.0, so that other CSAs can use the model to produce Scrip for their operations.

Inquiries to CSA Works are best made from December through March; during the growing season, the staff is working in the field.

CSA Works
115 Bay St.
Hadley, MA 01035
Phone: 413-586-5133

Just Food is an alliance of food, farming, anti-hunger, urban gardening, consumer, and community development groups working together for a more just and sustainable food system. The Just Food alliance promoted community-based solutions to food, farming and hunger problems in New York City, and is a model other urban areas may wish to explore. They helped establish New York City consumer bases for six rural CSAs in 1996, and planned to work with five more in 1997.

Just Food
290 Riverside Drive #15D
New York, NY 10025-5287
Phone/fax: 212-666-2168

Land Tenure and Land Trusts for CSAs

(See *Appendix B*)

Resources for Congregation Supported Agriculture

The Land Stewardship Project (LSP) offers several resources for those interested in congregation supported agriculture:
To Till and Keep It: New Models for Congregational Involvement With the Land is a booklet containing more than fifty pages of tips and real-life anecdotes about how congregations can become more involved in a locally-based, sustainable food production system. Written by CSA pioneer Dan Guenthner of Osceola, WI, this guide outlines how community supported farms are starting to hook up with congregations interested in establishing ties with the land and sustainable agriculture. To order, send five dollars ($5.33 for Minnesota residents) to the LSP.

The Land Stewardship Congregational Directory & Resource Guide is a nineteen page booklet that describes Minnesota congregations whose members support local growers and community gardens. The guide describes the LSP Congregational Network, summarizes various congregational land and food projects and offers several resources, including a list of area food pantries that accept surplus produce. To order, send three dollars ($3.20 for Minnesota residents) to the LSP.

Our Garden: A Project That Supports Our Community and Protects Our Land is a thirteen minute video that details how Redeemer Lutheran Church set up a garden that serves its neighbors in need. The congregation and staff of the church, in conjunction with the Land

Stewardship Project and other organizations, businesses and individuals, have since 1985 produced approximately 100,000 pounds of organically grown produce that has been distributed to area food pantries and senior citizens' centers in Winona, Minnesota. The video offers advice on how such a garden can be set up and run in a community. To order, send seventeen dollars ($17.98 for Minnesota residents) to the LSP.

The Land Stewardship Congregational Tool Kit contains videos, resource materials and activities for small and large group gatherings with a focus on building healthy communities by linking people with their food, the land and each other. To inquire about renting the kit, call the LSP.

Land Stewardship Project
2200 4th St.
White Bear Lake, MN 55110
Phone: 612-653-0618

Religious Congregations on the Land: The Practical Links Between Community, Sustainable Land Use, and Spiritual Charism, 1996, published by the National Catholic Rural Life Conference. The book is fifteen dollars, which incudes shipping and handling. Payment must accompany your order.

National Catholic Rural Life Conference
4625 Beaver Ave.
Des Moines, IA 50310-2199
Phone: 515-270-2634

General Resources for Farm Support and Preservation

Appropriate Technology Transfer for Rural Areas (ATTRA) is an organization that disseminates information about sustainable agriculture and low-input farming. ATTRA maintains a vast resource center, publishes a quarterly newsletter and updated lists of resource materials. ATTRA is a non-profit organization dedicated to helping communities and individuals find sustainable ways of improving the quality of life using the skills and resources at hand. It is funded by the USDA.
ATTRA
PO Box 3657
Fayetteville, AR 72702
Phone: 1-800-346-9140

The Farming Alternatives Program at Cornell University has been established since 1987 to promote a sustainable food and agriculture system which supports farm families and their communities. The program focus is research and education relating to community agriculture. They sponsor *Farming for the Future—A Leadership Initiative for Community Agriculture Development,* and in-service training and support. They also publish a newsletter.

Farming Alternatives Program
Warren Hall
Cornell University
Ithaca, NY 14853
Phone: 607-255-9832
E-mail: jmp32@cornell.edu

The Sustainable Agriculture Network is a national network of people from land grant universities, private non-profit groups, agribusiness, and extension services. It strives to promote effective, decentralized communication about sustainable agriculture. It's resources include a directory of experts, computer-linked discussion groups, and publishing help.

Sustainable Agricultural Network
c/o AFSIC, Room 304
National Agricultural Library
10301 Baltimore Blvd.
Beltsville, MD 20705
Phone: 301-504-6425

Alternative Farming Systems Information Center is part of the National Agricultural Library, the world's largest collection of information on farming and gardening. The Alternative section is set up as a reference service for anyone interested in alternative farming practices, and it can make complimentary computer searches of the vast AGRICOLA database. They furnish an *Annotated Bibliography and Resource Guide on Community Supported Agriculture,* refer experts and organizations, and lend books.

Alternative Farming Systems Information Center
National Agricultural Library
10301 Baltimore Blvd., Room 304
Beltsville, MD 20705
Phone: 301-504-6559

The **Center for Rural Affairs** was formed by rural Nebraskans concerned about the role of public policy in the decline of family farms and rural communities. A non-profit organization, it's purpose is to provoke public thought about social, economic and environmental issues affecting rural America. Publications include low-cost and free materials for beginning and family farmers, and a newsletter for beginning farmers. The Center engages in innovative projects, focused on topics such as emerging technologies, sustainable agriculture, rural economic policy, and Rural Community Development.

Center for Rural Affairs
PO Box 406
Walthill, NE 68067
Phone: 402-846-5428

The **International Alliance for Sustainable Agriculture** is a non-profit organization of individuals and groups cooperating to develop economically viable, ecologically sound, socially just, and humane agricultural systems around the world. It engages in research and documentation, network building, and education.

International Alliance for Sustainable Agriculture
1701 University Ave S.E. Room 202
Minneapolis, MN 55414

The **New England Small Farm Institute** is a small, non-profit organization dedicated to promoting the sustainable use of the region's agricultural resources. It manages 400 acres of public forest and farmland, works to refine viable small-farm management systems, and to provide information and training in ecologically responsible commercial production of food and fuel. The Institute has a substantial agricultural library, including a CSA collection and New England Land Link Directory.

New England Small Farm Institute
Box 937
Belchertown, MA 01007
Phone: 1-413-323-4531
E-mail: nesfi@igc.apc.org

Useful Publications

Community Supported Agriculture: Making the Connection is a 198-page handbook for producers that pulls together the experience of many innovative projects. While describing the diversity of CSA, it addresses common questions and concerns. In addition to the narrative text and examples from CSA farms across the country, the handbook includes simple model forms for use in running CSA projects. Charts for planning production offer handy information, and worksheets help farmers consider their own situations. The handbook also provides additional detail on such topics as legal issues of CSAs, writing newsletters, and post-harvest handling. To order send twenty-five dollars per copy, plus five dollars shipping and handling. Make checks payable to "UC Regents."

CSA Handbook
c/o UC Cooperative Extension
11477 E Avenue
Auburn, CA 95603
Phone : 1-916-889-7385
E-mail: ceplacer@ucdavis.edu

Rebirth of the Small Family Farm—A Handbook for Starting a Successful Organic Farm Based on the CSA Concept. Written by farmers Bob and Bonnie Gregson, this sixty-four page book tells how they started a thirty-eight share CSA and made it work for them and their neighbors. Available for $9.95 postpaid from:

IMF Associates
Box 2542
Vashon Island, Wa. 98070

The Farmland Preservation Directory is a sourcebook of organizations, models, and printed materials oriented toward helping preserve farms and farmland. It is available from the Natural Resources Defense Council.

Farmland Preservation Directory
Natural Resources Defense Council
122 East 42nd Street
New York, NY 10168
Phone: 212-949-0049

The CSA Food Book, by Elizabeth Henderson, David Stern and the Genesee Valley Organic CSA Project. This is an illustrated, 128-page loose-leaf document, which lists crops and appropriate recipes seasonally. Each crop is described, giving pertinent historical and nutritional information, as well as storage tips. The guide is expandable with your own notes. This handbook was composed in response to questions from CSA members on how to use the variety of food in their weekly shares. Profits from the *CSA Food Book* go towards scholarships to enable low-income people to join the CSA.

A single copy is nine dollars; unbound copies are eight dollars. CSAs may reprint copies for members if they pay Rose Valley $2.50 per copy in royalties. The *CSA Food Book* may be ordered from:

Elizabeth Henderson
Rose Valley Farm
PO Box 149
Rose Valley, New York, 14542-0149.

CSA Farm Network, by Steve Gilman, includes many articles on CSA, resource listings, and a list of CSAs in the Northeast. Published in 1996 and 1997, with a planned edition for 1998 by the Northeast Organic Farming Association, with the support of the Northeast Sustainable Agriculture Research and Education Program of USDA. Contact Steve Gilman, coordinator of CSA Net, 130 Ruckytucks Rd., Stillwater, NY 12170.

Growing for Market, a non-glossy magazine for small-scale farmers. CSA is a major topic, including profiles of CSA farms, articles on organizational aspects of CSA, and practical information about organic food and flower production. Twenty-seven dollars per year from:

Growing for Market
PO Box 3747
Lawrence, KS 66046
Phone: 913-841-2559

A List of Buying Clubs by Region is available by sending a self-addressed, stamped envelope to Co-op Directory Services, c/o Kris Olsen, 919 21st Avenue South, Minneapolis, MN 55404.

Threefold Review is a journal edited by Gary Lamb of the Hawthorne Valley Farm. The journal frequently publishes articles on various aspects of associative economy, an economic approach at the heart of many CSAs. Write to the *Threefold Review,* PO Box 6, Philmont, NY 12565.

Books for CSA Homemakers

Louise's Leaves is a paperback cook's journal about the crops typically grown by CSAs. It is arranged chronologically—from the start of the growing season when spinach and asparagus are plentiful, to the winter when beans, sweet potatoes, and squash are being drawn from storage for meals. Each "leaf,' or page, contains cultural and nutritional information about one particular crop, and suggestions on preparation. By Louise Frazier, 1994, ($15.95). A related chart is also available, *Louise's Complementary Herbs & Spices.* This laminated chart of which herbs and spices go naturally with each crop from the garden is available at a cost of five dollars. Both published by:

Biodynamic Farming and Gardening Association
PO Box 550
Kimberton, PA 19442
Phone: 800-516-7797

From Asparagus to Zucchini is a valuable 195-page resource for households participating in a CSA. With an alphabetical recipe and information section for each vegetable, this book details ways to use and preserve seasonal produce. It was designed to address the concerns expressed by consumers as they struggle to overcome "supermarket withdrawal," seasonal vegetable fluctuations, onslaughts of unknown vegetables, root crops tedium, and other typical consumer challenges. Mixed in with the recipes are educational tidbits about CSA and the larger food and agricultural system and descriptions of each local farm. The book has been an effective fund-raiser for the sixteen farms in the Madison Area CSA Coalition. For price and other information, contact:

MASCAC (Madison Area Community Supported Agriculture Coalition)
c/o Wisconsin Rural Development Center
125 Brookwood Drive
Mount Horeb, WI 53572
Phone: 608-437-5971

CSA on the Internet

CSA resources on the Internet are constantly changing and improving. Because things change so much, the most effective way to get a sense of what is currently available is to use a World Wide Web (WWW) search engine such as Alta Vista or Yahoo. Just enter "CSA" or "Community Supported Agriculture" as the keyword in the search box, and start the engine on its quest for related homepages. You will be rewarded.

If you or members of your CSA do not have direct access to the Internet, you may wish to check public libraries and schools. Many have Internet connections, and some libraries will even help you with the search if you are new to the Internet.

The e-mail list "csa-l@prairienet.org" is dedicated to networking on Community Supported Agriculture. The list is free to all who wish to subscribe, and includes discussions on goals, distribution styles, outreach tools, member retention, educational work, connections with non-CSA organizations, and related topics. From time to time it also includes the exchange of newsletter articles and information about resources on CSA such as networking organizations, books, journals, videos, audio tapes from conferences, speaker tours of interest to CSAs, and so forth. The intention of the mailing list is to take advantage of the collective knowledge of as many CSAs as possible to help the CSA movement grow.

The list owners are John Barclay (jbarclay@prairienet.org) of Prairieland CSA in Champaign, Illinois and Sarah Milstein (milstein@pipeline.com) of Roxbury Farm CSA in New York.

A related page on the Web, "http://www.prairienet.org/", archives the messages posted to the CSA-L mailing list, and also holds biographies of CSA-L participants.

Farms Profiled in this Book

Brookfield Farm CSA
Dan Kaplan
24 Hulst Rd.
Amherst, MA 01002

Caretaker Farm
Sam and Elizabeth Smith
1210 Hancock Road
Williamstown, MA 01267-3026

CSA Garden at Great Barrington
(Now the Mahaiwe Harvest CSA)
David Inglis
342 North Plain Rd.
Housatonic, MA 01236-9741

Community Supported Composting
Woods End Agricultural Institute (William Brinton)
Old Rome Rd., Route 2, Box 1850
Mt. Vernon, ME 04352

Fairfield Gardens
Michael Abelman
598 North Fairview Ave.
Goleta, CA 93117

Food Bank Farm
Michael Docter and Linda Hildebrand
115 Bay Rd.
Hadley, MA 01035

Forty Acres and Ewe
Jim Bruns and Donna Goodlaxon
339 10th St.
Prairie Farm, WI 54762

Dartmouth Organic Farm
Dartmouth College
PO Box 9
Hanover, NH 03755

Good Humus Produce CSA
12255 County Rd. #84A
Capay, CA 95607

Hawthorne Valley Farm
Steffen and Rachel Schneider, and Tom Meyers
Box 225A, RD 2
Ghent, NY 12075

Homeless Garden Project
PO Box 617
Santa Cruz, CA 95061

Kimberton CSA
Kerry and Barbara Sullivan
PO Box 192
Kimberton, PA 19442

Madison Area Community Supported Agriculture (MACSAC)
c/o Wisconsin Rural Development Center
125 Brookwood Drive
Mount Horeb, WI 53572
(A coalition of sixteen CSA farms as of 1996.)

Minnesota and Western Wisconsin Community Supported Farms
c/o The Land Stewardship Project
2200 4th St.
White Bear Lake, MN 55110
(A coalition of twenty-six CSA farms as of 1996.)

Marian Trading Co.
Gena Nonini
PO Box 9176
Fresno, CA 93780-9176

Natick Community Organic Farm
117 Eliot St.
South Natick, MA, 01760

Roxbury Farm CSA
c/o Jean-Paul Courtens
124 Roxbury Rd.
Hudson, NY 12534

Temple-Wilton Community Farm
c/o Anthony Graham
135 Temple Road
West Wilton, NH 03086

Supplies And Support Tools

Biodynamic Preparations
The Josephine Porter Institute for Applied Biodynamics, Inc.
PO Box 133
Woolwine, VA 24185-0133
Phone: 540-930-2463

Certification of Biodynamic® Farms or Gardens
The Demeter Association, Inc.
Britt Road
Aurora, NY 13026
Phone: 315-364-5616

Bibliography

Ableman, Michael. *From the Good Earth: A Celebration of Growing Food Around the World.* New York: Harry N. Abrams, 1993.

Bear, Firman. *Cation and Anion Relationship and their Bearing on Crop Quality.* New Brunswick, NJ: Rutgers University, Department of Soils, 1947.

———. "Soil and Fertilizers." In *Earth the Stuff of Life.* Norman, Okla.: University of Oklahoma Press, 1986.

Berry, Wendell. *Sex, Economy, Freedom & Community: Eight Essays.* New York: Pantheon Books, 1994.

———. *The Unsettling of America: Culture and Agricuture.* San Francisco: Sierra Club Books, 1977.

Brinton, William F. "Community Supported Composting," *Biodynamics* 194, (July/August, 1994). Kimberton, Pa.: Biodynamic Farming and Gardening Association.

Brown, Lester R. *Tough Choices: Facing the Challenge of Food Scarcity.* Washington: Worldwatch Institute, 1996.

Brown, Lester, and Hal Kane. *Full House: Reassessing the Earth's Population Carrying Capacity.* Washington: Worldwatch Institute, 1994.

Bruell, Dieter. *Waldorf School and Three-fold Structure.* Nearchus, Holland: Lazarus Publishing, 1995. (Soon to be published in the U.S.)

Centre des Jeunes Dirigeants d'Enterprise. "The Company in the 21st Century." *International Herald Tribune,* 10/15/96.

Derry, Evelyn Francis. *The Christian Year.* London: The Christian Community Press, 1967.

Douglas, William Campbell, M.D. *The Milk Book* (originally published as *The Milk of Human Kindness Is Not Pastuerized*). Atlanta: Second Opinion Publishing, 1996.

Kervan, C.L. *Biological Transmutation.* London: Crosby-Leonard Publishing, 1972.

Peck, Scott. *A Different Drum: Community Building and Placemaking.* Touchstone Books.

Pfeiffer, Ehrenfried. *Soil Fertility.* East Grinstead, England: 1983.

Remer, Nicholas. *Laws of Life in Agriculture.* Kimberton, Pa.: Biodynamic Farming and Gardening Association, 1995.

————. *Organischer Dunger.* Amelinghausen: 1980. (Published in an English translation as *Organic Manure* by Mercury Press, Spring Valley, N.Y.).

Robbins, John. *Diet for a New America.* Walpole, N.H.: Stillpoint Publishing: 1990.

Schmidt, Gerhard, M.D. *The Dynamics of Nutrition.* Kimberton, Pa.: Biodynamic Farming and Gardening Association, 1980.

————. *The Essentials of Nutrition.* Kimberton, Pa.: Biodynamic Farming and Gardening Association, 1987.

Soils and Men—Yearbook of Agriculture 1938. United States Department of Agriculture.

Steiner, Rudolf. *Agriculture.* Kimberton, Pa.: The Biodynamic Farming and Gardening Association, 1993.

————. *Towards Social Renewal.* Hudson, N.Y.: Anthroposophic Press, 1993.

————. *World Economy: The Formation of a Science of World Economics.* London: Rudolf Steiner Press, 1977.

Stella*Natura: Kimberton Hills Agricultural Calendar. Kimberton, Pa.: Biodynamic Farming and Gardening Association.

Thunn, Maria and Mathias Thunn. *Working with the Stars.* Launceston, Cornwall, England: Lanthorn Press. (Published annually.)

Voronkow, Zelchan, and Lukewitz. *Silicium und Leben.* Berlin: 1975.

Welsh, Rick. *The Industrial Reorganization of U.S. Agriculture: An Overview and Background Report.* Greenbelt, Md.: Henry A. Wallace Institute for Alternative Agriculture, 1996.

About The Authors

Trauger Groh has been a farmer for forty years and has been at the leading edge of the organic, biodynamic and community farm movements. After helping to establish a widely known community-supported farm in North Germany, he settled down in Wilton, NH, where he helped to start the Temple-Wilton Community Farm. This farm raises the fruits, vegetables, milk, and eggs for about one hundred New Hampshire families. Trauger is also active as a consultant for many other farmers and farming groups in America and abroad. Well known in the international agricultural community, he has lectured hundreds of times on social, educational and agricultural issues. In recent years, on regular visits to Russia he has been helping to build up low-input organic farms, and presenting lectures on farm-related issues.

At present, Trauger is particularly active through The Biodynamic Farmers of the Northeast, and the Biodynamic Association of America. Locally, he is active through the Cadmus Corporation, a non-profit organization dedicated to acquiring land and the other resources necessary to enable farmers to farm, and for training and research. Cadmus is dedicated to finding practical ways to help develop the farms of tomorrow. Trauger is actively raising funds to support this endeavor. You may contact him through the Cadmus Corporation, PO Box 333, Wilton, NH 03086.

Steven McFadden lives about five miles down the road from Trauger. An independent journalist since 1975, Steven has published hundreds of magazine and newspaper articles, and is the author of several books, including:

Profiles in Wisdom: Native Elders Speak About the Earth. Bear & Co., 1990.

Ancient Voices, Current Affairs: The Legend of the Rainbow Warriors. Bear & Co., 1991.

The Little Book of Native American Wisdom. Element Books, 1994.

Teach Us to Number Our Days. Element Books, 1995.

Steven is founder and director of The Wisdom Conservancy at Merriam Hill Education Center, a private, non-profit institute. The mission of the institute is to conserve, communicate, and encourage wisdom via modern media, and thereby to support and encourage the public in cultivating and applying wisdom. He travels widely to study, and to present lectures and workshops. You may contact him through The Wisdom Conservancy at Merriam Hill Education Center, 148 Merriam Hill Rd., Greenville, NH 03048.

Trauger and Steven's book, *Farms of Tomorrow* (1990) has been translated and published in Russian, Korean, and Japanese.